The Excitation Functior

CW00471177

Mohsen Challan

The Excitation Functions for Some Cyclotron Produced Radionuclides

the power horse among the radionuclides used in nuclear medicine

LAP LAMBERT Academic Publishing

Imprint

Cover image: www.ingimage.com

Publisher:
LAP LAMBERT Academic Publishing
is a trademark of
Dodo Books Indian Ocean Ltd. and OmniScriptum S.R.L publishing group

120 High Road, East Finchley, London, N2 9ED, United Kingdom
Str. Armeneasca 28/1, office 1, Chisinau MD-2012, Republic of Moldova, Europe
Printed at: see last page
ISBN: 978-3-659-62835-1

Zugl. / Approved by: Mansoura, Mansoura University, Diss. 2006

STUDY OF THE EXCITATION FUNCTIONS FOR SOME CYCLOTRON PRODUCED RADIONUCLIDES

Presented by

Mohsen Bekheet Mohamed
Lecturer Assistant
Experimental Nuclear Physics Department (Cyclotron Project)
Nuclear Research Center – Atomic Energy Authority

Supervisors

Prof. Dr. M. A. Abou-Zeid
Prof. of Nuclear Physics – Physics Department
Faculty of Science – Mansoura University

Prof. Dr. M. N. H. Comsan
Prof. of Nuclear Physics – Experimental Nuclear Physics Department
Nuclear Research Center – Atomic Energy Authority

CONTENTS

Acknowledgment
Abstract
List of Figures
List of Tables

Chapter I

General Introduction

1-1	Introduction	1
1-2	Reaction Q-Values, and Thresholds	2
1-3	Status of Available Nuclear Data	5
1-4	Nuclear Data Needs and Compilation	7
1-5	Cyclotron Production of Medically Used Radionuclides	9
	1-5.1 Target Chemistry and Targetry	13
	1-5.2 High Current Irradiation at Cyclotrons	14
1-6	Calculation of the Activation Cross-Section	16
	1-6.1 Cumulative Cross-Sections	21
	1-6.2 Isomeric Cross Section Ratios	23
1-7	Yield, and Effective Cross Section	24
	1-7.1 Production Cross-Section	25
	1-7.2 Production Yield	25
	1-7.3 Thin Target Yield, Effective Cross Section	28
	1-7.4 Thick Target Yield Calculation	29
1-8	The Aim of the Present Study	31

Chapter II

Literature Review

2-1	Literature Review	33
	2-1.1 Technetium-99m	33
	2-1.2 Technetium-99m Analogues	37
	2-1.3 Terbium	38
2-2	Summary of Earlier Investigations	40

Chapter III

Experimental Techniques and Measurements

3-1 Introduction 43
3-2 The Cyclotron 43
3-3 Stacked Foils Technique 48
3-4 Calibration of Cyclotron Beam Energy 50
3-5 Beam Intensity Monitoring 53
3-6 Associated Counting Electronics 54
3-6.1 γ–Spectrometer 54
3-6.2 HPGe-Detector Efficiency 56
3-7 Analysis of Gamma Spectra 60
3-7.1 Nuclide Identification 63
3-7.2 Selection of Reliable Data 67
3-7.3 Definition of Detection Probability 67
3-8 The Sources of Uncertainties 69
3-8.1 Uncertainties in Proton Energies, Beam Fluctuation 69
3-8.2 Uncertainties in Net Peak Area, Decay Data 70
3-8.3 Uncertainties in the Absolute Efficiency, Time Factor, and Dead Time 71
3-8.4 Uncertainties in Irradiated Nuclei, Impurities, Recoil Contamination 71
3-8.5 Uncertainties due to γ-Interference, γ-Self-Absorption, and γ-Attenuation 72
3-8.6 Uncertainties due to Interfering Processes 73
3-8.7 Uncertainties due to Random Coincidence Summing 76
3-8.8 Uncertainties due to True Coincidence 80
3-8.8.1 γ–γ Coincidence for Extended Sources 82
3-8.8.2 The Volume Effect Factor F_V 85
3-8.8.3 Peak Area Correction 88
3-8.9 Uncertainties of Fluence Response 91
3-8.9.1 Neutron-Production Probability 94
3-8.9.2 Neutron Transmission Probability 97
3-9 Estimated Correction Values for Uncertainty Sources 102

Chapter IV

Nuclear Model Calculations

4-1 Introduction 105
4-2 Outline of Nuclear Reaction Mechanisms 106
4-3 Description of Code ALICE-91 106
4-4 Description of Code EMPIRE-II 109
4-4.1 Multi-Step Direct Model 111
4-4.2 Multi-Step Compound Model 112
4-4.3 Coupling Between MSC, and MSD 114
4-4.4 Exciton Model 115
4-4.5 Monte Carlo Preequilibrium Model 117
4-4.6 Compound Nucleus Model 120
4-4.7 Level Densities 121
4-4.7.1 Gilbert-Cameron Approach 122
4-4.7.2 Dynamic Approach 123
4-4.7.3 Hartree-Fock-BCS Approach 124
4-4.8 Width Fluctuation Correction 125
4-4.9 Flow of EMPIRE-II Calculations 127
4-4.10 RIPL-2 Input Parameters Libraries 129
4-4.11 Input and Output Files 131
4-4.12 Models Compatibility and Priorities 131

Chapter V

Results and Discussion

5-1 Calculation of the Theoretical Cross-Section 133
5-1.1 Analysis with Code ALICE-91 133
5-1.2 Analysis with Code EMPIRE-II 134
5-1.3 Discrete Level Cross Sections 136
5-2 The Excitation Function for Produced Radionuclides from Mo Targets 142
5-2.1 The $^{nat}Mo(p,x)^{92m}Nb$ Reaction 143
5-2.2 The $^{nat}Mo(p,x)^{94g}Tc$ Reaction 143
5-2.3 The $^{nat}Mo(p,x)^{95g}Tc$ Reaction 146
5-2.4 The $^{nat}Mo(p,x)^{95m}Tc$ Reaction 148
5-2.5 The $^{nat}Mo(p,x)^{96m+g}Tc$ Reaction 150
5-2.6 The $^{nat}Mo(p,x)^{99m}Tc$ Reaction 152
5-3 The Excitation Function for Produced Radionuclides from Gd Targets 155
5-3.1 The $^{nat}Gd(p,x)^{152m+g}Tb$ Reaction 155

5-3.2 The $^{nat}Gd(p,x)^{154g}Tb$ Reaction 157
5-3.3 The $^{nat}Gd(p,x)^{154m}Tb$ Reaction 159
5-3.4 The $^{nat}Gd(p,x)^{155}Tb$ Reaction 161
5-3.5 The $^{nat}Gd(p,x)^{156}Tb$ Reaction 161
5-3.6 The $^{nat}Gd(p,x)^{160}Tb$ Reaction 164
5-4 Differential, and Integral Yields for Mo, and Gd Targets 166
5-5 Calibration Curves for Thin Layer Activation Technique 172
5-6 Discussion 175
5-6.1 Influence of Interfering Processes 175
5-6.2 Trends in Integral Cross Sections 176
5-6.3 Prospective for Fluence Response Modelling 178
5-7 Conclusion 179
References

ACKNOWLEDGMENT

Thanks to ALLAH who enabled me to perform this work

The author wishes to give his gratitude, appreciation, sincere thanks to:

Prof. Dr. Mohamed N. H. Comsan (supervisor) Professor of Nuclear Physics, Experimental Nuclear Physics Department, Atomic Energy Authority, for his scientific support throughout this study. Prof. Comsan was my supervisor during my M.Sc. degree work and I enjoyed his kindness and encouragement. I thank him for his keeping the study plane achievable, and steady interest, encouragement feelings, progressive discussion. He helped me very much to accelerate procedure of the Ph.D. work.

Prof. Dr. Mahmoud A. M. Abou-Zeid (supervisor) Professor of Nuclear Physics, Faculty of Science, Mansoura University for his kind supervision during this thesis, Prof. Abou-Zeid was my Supervisor during my M.Sc. degree work and I enjoyed his kindness and encouragement. I thank him for the fruitful discussion and for his interest. Also my great thanks to his nuclear physics group for their interest, all the generous efforts, valuable suggestions, and continuos encouragement.

The Author gratefully acknowledges Prof. Dr. Mohammed A. Madcore the Head of Physics Department, Faculty of Science, Mansoura University, and Prof. Dr. Shokry M. Saad the Supervisor of Cyclotron Project for their kind help. My deep thanks for Prof. Dr. Zeinab Abdu Saleh for here usual helpful, kind support, and Prof. Dr. M. A. Ali for his kind advice. I am also indebted to Prof. Dr. Samia El-Gibely the Head of Experimental Nuclear Physics Department, NRC, Atomic Energy Authority for her valuable help.

The Author deeply thanks Demonstrator Gamal S. Moawad, Dr. Eng. Adel L. Ismail, Dr. Eng. Mohamed N. El-Shazly, and Mr. Nasr Abd-Alrahman, for their valuable help. My appreciation to the Technical, and Operation Group of the Inshas Cyclotron Facility, Department of Experimental Nuclear Physics, especially my friend, and colleague M. Sc. Megahed Al-Abyad, for their kind helps in various aspects of experiment preparation and data collection.

Abstract

Measurement of reaction cross sections is one of the main aims of many experimentalists due to the need of proven cross section data for the reactions. Moreover, the lack and discrepancies among the available data are still existing. Our concern is focused on targets that can produce radionuclides of main use in nuclear medicine and thin layer activation techniques.

Excitation functions of some Molybdenum and Gadolinium isotopes were measured using the gamma spectra from the (p,x) reactions of high purity natural targets. These reactions could be produced by means of relatively low projectile energies.

We measured the excitation functions of (^{92,94,95,96,97,98,100}Mo, and ^{152,154,155,156,157,158,160}Gd) reactions induced by protons in the energy range from 5 to 18 MeV.

The measured cross sections have been carried out by applying the activation technique using a HPGe detector as a γ–ray spectrometer. So, we were planing to produce a reliable data with the help of multi monitor simultaneous measurements such as the ^{nat}Ti(p,x)^{48}V, ^{nat}Cu(p,x)^{62}Zn, ^{nat}Cu(p,x)^{63}Zn, ^{nat}Cu(p,x)^{65}Zn. This study was performed at the compact variable energy cyclotron (VEC), model MGC-20 of the Inshas Cyclotron Facility, AEA, Cairo, Egypt.

Since, ^{99m}Tc is the power horse among the radionuclides used in nuclear medicine, special attention was devoted to its production. Not only ^{99m}Tc but also some other isotopes such as ^{94}Tc, ^{95m}Tc, ^{95g}Tc, ^{96}Tc could be produced from Mo via charged particle induced reactions with its related quality and security. We have also chosen the rare earth element Gadolinium as a target to produce Terbium radionuclides (^{152,154m,154g,155,156,160}Tb) of deficient nuclear data. These radionuclides could be promising for their radiation characters that nominate its uses in nuclear medicine, especially $^{152m+g}$Tb, ^{154m}Tb in magnetic resonance imaging, ^{155,156}Tb, ^{160}Tb could be used in diagnosis, and therapy, respectively.

We surveyed the existing data for ^{99m}Tc production, and found that for some works, the results are promising, while the results obtained from some others are discrepant. So we here look for a new aspects. These aspects are looking for the effects of presence of secondary beams that could produce the same reactions, while these particles propagate through the stack targets.

The present study pays attention to investigate the contribution of secondary beams. The consideration of presence the secondary beams enable us to verify this prospective, which providing us with reliable reaction cross sections, a substantial understand information on the investigated processes. The data obtained from these measurements could be of benefit for routine production of concered radionuclides.

We aimed to investigate to how extent these secondary reactions may contribute or not to the essential production cross sections for these radionuclides. The main induced secondary reactions originate from propagating protons and neutrons. From the obtained data we calculated the integrated yield to investigate the possibility of using these reactions in different applications technique such as radionuclides production routes, and thin layer activation technique. The yield profile enables us to deduce some criteria in the production process.

Theoretical model calculations using the EMPIRE-II and the ALICE-91 codes were undertaken to describe the cross sections of all induced reactions. In most of the cases, the calculated values fit fairly well with the experimental results.

List of Figures

Figure	Description	Page
Fig. (1-1):	Scheme of Disciplines and Activities Involved in a Cyclotron Facility.	10
Fig. (3-1):	Picture for the 103 AVF (MGC-20) Cyclotron (Level-II) at the Inshas Cyclotron Facility.	45
Fig. (3-2):	Layout of Cyclotron and Beam Transport System at the Cyclotron Laboratory of Inshas MGC-20 (AVF).	46
Fig. (3-3):	Stacked Foils Technique Arrangement.	50
Fig. (3-4):	Indicates the Agreement between Recommended Cross Section Variation for Monitor Reactions and the Corresponding Experimental Data of this Work.	52
Fig. (3-5):	Gamma Ray Spectrometer.	55
Fig. (3-6):	Sample of Gamma Ray Spectrum of Natural Copper Sample Irradiated with Protons for Beam Intensity Monitoring.	56
Fig. (3-7):	Sample of Gamma Ray Spectrum of Natural Titanium Sample Irradiated with Protons for Beam Intensity Monitoring.	57
Fig. (3-8):	Absolute Full Energy Peak Efficiency of HPGe-Detector, Experimental Points, and Fitted Curves at Different Distance.	58
Fig. (3-9):	Evaluation of one Spectrum Peak by Using Gaussian Fit, and Linear Background Subtraction	59
Fig. (3-10):	Sample of Gamma Ray Spectra of Natural Molybdenum Sample Irradiated with Protons for Reaction Cross Section Measurements one hour after EOB.	61
Fig. (3-11):	Sample of Gamma Ray Spectra of Natural Molybdenum Sample Irradiated with Protons for Reaction Cross Section Measurements 19 day after EOB.	61
Fig. (3-12):	Sample of Gamma Ray Spectra of Natural Gadolinium Sample Irradiated with Protons for Reaction Cross Section Measurements one hour after EOB.	62
Fig. (3-13):	Sample of Gamma Ray Spectra of Natural Gadolinium Sample Irradiated with Protons for Reaction Cross Section Measurements 10 day after EOB.	62
Fig. (3-14):	The Decay Scheme of ^{96}Tc	65
Fig. (3-15):	The Decay Scheme of ^{99m}Tc	65
Fig. (3-16):	The Decay Scheme of ^{152}Tb	66

Fig. (3-17):	The Decay Scheme of ^{160}Tb	66
Fig. (3-18):	A Simple Decay Scheme Showing (a) True-Coincidence Summing A=B+C. (b) True-Coincidence Loss $\gamma-\gamma$ and γ–KX (IC), and (c) True-Coincidence Loss $\gamma-\gamma$ and γ–KX (EC) Effects.	81
Fig. (3-19):	Neutron Production Cross Section from Proton Induced Reaction for Natural Gadolinium, and Molybdenum Target Foils as a Function of the Proton Energy, E_p up to 20 MeV, ALICE-91 Code.	93
Fig. (3-20):	Differential Cross-Section Neutron Spectrum from Natural Gadolinium, and Molybdenum Target Foils Bombarded with 20 MeV Protons, ALICE-91 Code.	93
Fig. (3-21):	Neutron Production Cross Section from Neutron Induced Reaction for Natural Gadolinium, and Molybdenum Target Foils as a Function of the Neutron Energy, E_n up to 5 MeV, Using ALICE-91 Code.	96
Fig. (3-22):	Neutron Induces Capture Cross Section for Natural Gadolinium, and Molybdenum Target Foils as a Function of the Neutron Energy, E_n up to 5 MeV in Log-Log Scale.	97
Fig. (3-23):	Relative Values of the Neutron Production Probability, $f^{(1)}(E_i,t)$, in Arbitrary Units, for Gadolinium, and Molybdenum Target Foils of Thickness t= 10, and 25 μm, respectively as a Function of the Proton Energy, E_p (MeV).	97
Fig. (3-24):	Relative Values of the Ejected Neutron Transmission Probability, $f^{(2)}(E_i,t)$, in Arbitrary Units, for Gadolinium, and Molybdenum Target Foils as a Function of the Ejected Neutron Energy, E_n (MeV).	99
Fig. (3-25):	Relative Values of the Overall Proton Fluence Response Probability per unit Proton Energy, $F(E_i,t)/E_p$, in Arbitrary Units, for Gadolinium, and Molybdenum Target Foils, as a Function of the Incident Proton Energy, E_p (MeV).	101
Fig. (4-1):	Flow Chart of EMPIRE-II Calculations	131
Fig. (5-1):	The Excitation Function of the $^{nat}Mo(p,x)^{92m}Nb$ Reaction.	144
Fig. (5-2):	The Excitation Function of the $^{nat}Mo(p,x)^{94g}Tc$ Reaction.	145
Fig. (5-3):	The Excitation Function of the $^{nat}Mo(p,x)^{95g}Tc$ Reaction.	147

Fig. (5-4):	The Excitation Function of the $^{nat}Mo(p,x)^{95m}Tc$ Reaction.	148
Fig. (5-5):	The Isomeric Cross Section Ratios Values for $^{95g}Tc/^{95m}Tc$.	149
Fig. (5-6):	The Excitation Function of the $^{nat}Mo(p,x)^{96mg}Tc$ Reaction.	151
Fig. (5-7):	The Excitation Function of the $^{nat}Mo(p,x)^{99m}Tc$ Reaction.	154
Fig. (5-8):	The Excitation Function of the $^{nat}Gd(p,x)^{152g}Tb$ Reaction.	156
Fig. (5-9):	The Excitation Function of the $^{nat}Gd(p,x)^{154g}Tb$ Reaction.	158
Fig. (5-10):	The Excitation Function of the $^{nat}Gd(p,x)^{154m}Tb$ Reaction.	159
Fig. (5-11):	The Isomeric Cross Section Ratios Values for $^{154g}Tb/^{154m}Tb$.	160
Fig. (5-12):	The Excitation Function of the $^{nat}Gd(p,x)^{155}Tb$ Reaction.	162
Fig. (5-13):	The Excitation Function of the $^{nat}Gd(p,x)^{156}Tb$ Reaction.	163
Fig. (5-14):	The Excitation Function of the $^{nat}Gd(p,x)^{160}Tb$ Reaction.	165
Fig. (5-15):	The Differential Thin Targets Yields for the Production of ^{92m}Nb, ^{94}Tc, ^{95m}Tc, ^{95g}Tc, ^{96mg}Tc, and ^{99m}Tc via the $^{nat}Mo(p,x)$ Reactions.	166
Fig. (5-16):	The Differential Thin Targets Yields for the Production of ^{152g}Tb, ^{154m}Tb, ^{154g}Tb, ^{155}Tb, ^{156}Tb, and ^{160}Tb via the $^{nat}Gd(p,x)$ Reactions.	167
Fig. (5-17):	The Integral Thick Target Yield as a Function of Target Thickness (in MeV) for the Production of ^{92m}Nb.	168
Fig. (5-18):	The Integral Thick Target Yield as a Function of Target Thickness (in MeV) for the Production of ^{94g}Tc.	168
Fig. (5-19):	The Integral Thick Target Yield as a Function of Target Thickness (in MeV) for the Production of ^{95g}Tc.	168
Fig. (5-20):	The Integral Thick Target Yield as a Function of Target Thickness (in MeV) for the Production of ^{95m}Tc.	169

Fig. (5-21):	The Integral Thick Target Yield as a Function of Target Thickness (in MeV) for the Production of ^{96mg}Tc.	169
Fig. (5-22):	The Integral Thick Target Yield as a Function of Target Thickness (in MeV) for the Production of ^{99m}Tc.	169
Fig. (5-23):	The Integral Thick Target Yield as a Function of Target Thickness (in MeV) for the Production of ^{152g}Tb.	170
Fig. (5-24):	The Integral Thick Target Yield as a Function of Target Thickness (in MeV) for the Production of ^{154g}Tb.	170
Fig. (5-25):	The Integral Thick Target Yield as a Function of Target Thickness (in MeV) for the Production of ^{154m}Tb.	170
Fig. (5-26):	The Integral Thick Target Yield as a Function of Target Thickness (in MeV) for the Production of ^{155}Tb.	171
Fig. (5-27):	The Integral Thick Target Yields as a Function of Target Thickness (in MeV) for the Production of ^{156}Tb.	171
Fig. (5-28):	The Integral Thick Target Yield as a Function of Target Thickness (in MeV) for the Production of ^{160}Tb.	171
Fig. (5-29):	Presents the Integral Thick Target Yield for $^{nat}Mo(p,x)^{96}Tc$ Reaction in Comparison with the other Literature Data.	173
Fig. (5-30):	The Activity Distribution for $^{nat}Mo(p,x)^{96}Tc$ Reaction as a Function of the Penetrating Depth of the 18, 14, and 10 MeV Bombarding Proton Beam Calculated for Mo Using the Fitted Cross Sections of this Work in Comparison with other Literature Data.	174

List of Tables

Table	Description	Page
Table (1-1):	List of the Classified In-Vivo Radionuclides According to Their Chemical Behavior.	6
Table (1-2):	Potential of a Small Four Particles Cyclotron with Regard to Medical Radionuclides Production.	6
Table (1-3):	Types of Nuclear Data, and their Main Applications	8

Table (1-4):	Presents a list of some Important Positron Emitting Radionuclides Used for PET, their Half-Lives and Typical Nuclear Reactions for their Production Are Given.	11
Table (1-5):	Presents a list of some Important PET, and SPECT Radionuclides, and a Classification of Cyclotrons Used in their Production.	12
Table (1-6):	Characteristics of the Direct Detection of the Emitted Particles, and Activation Technique.	18
Table (2-1):	Principal Direct Nuclear Reactions, Decay Data Leading to the Main Radionuclides we are Interested in, that Occur in Proton Irradiated ^{nat}Mo and their Calculated Threshold Energies, E_{th}.	36
Table (2-2):	Principal Nuclear Reactions, Decay Data Leading to the Main Radionuclides We Are Interested in, that Occur in Proton Irradiated ^{nat}Gd and their Calculated Threshold Energies, E_{th}.	40
Table (2-3):	Gives a Summary of the Earlier Investigations of the Proton Induced Data from the Available Literature.	41
Table (3-1):	Particles Species, Their Energies and Intensities for the External Beam of the Inshas Cyclotron MGC-20.	44
Table (3-2):	Gives a Relation Between the Machine and the Particles, Particle Energy, Available Current.	47
Table (3-3):	Isotopic Composition of Natural Molybdenum, and Gadolinium Used in this Study.	49
Table (3-4):	Gamma-Ray Standard Sources Used in Efficiency Determination.	59
Table (3-5):	Parameters Used to Fit the Efficiency Function.	60
Table (3-6):	Formulae Used for Numerical Analysis of Spectra Peaks.	64
Table (3-7):	Major Uncertainty Sources and their Estimated Values in the Cross Section Measurements.	103
Table (5-1):	The Resultant Output Cross Section Values from Several Proton Induced Reactions on Molybdenum Targets, The Calculation of the Total Cross Section from the Individual Contributing Reactions Have Been Performed by Using ALICE-91 Code.	138
Table (5-2):	The Resultant Output Cross Section Values from Several Proton Induced Reactions on Molybdenum Targets, The Calculation of the Total Cross Section from the Individual Contributing Reactions Have Been Performed by Using EMPIRE-II Code.	139

Table (5-3):	The Resultant Output Cross Section Values from Several Proton Induced Reactions on Gadolinium Targets, The Calculation of the Total Cross Section from the Individual Contributing Reactions Have Been Performed by Using ALICE-91 Code.	140
Table (5-4):	The Resultant Output Cross Section Values from Several Proton Induced Reactions on Gadolinium Targets, The Calculation of the Total Cross Section from the Individual Contributing Reactions Have Been Performed by Using EMPIRE-II Code.	141
Table (5-5):	Measured Cross Sections Values for Production of ^{92m}Nb, on Natural Mo Targets at Different Energies.	144
Table (5-6):	Measured Cross Sections Values for Production of ^{94}Tc, on Natural Mo Targets at Different Energies.	145
Table (5-7):	Measured Cross Sections Values for Production of ^{95g}Tc, on Natural Mo Targets at Different Energies.	147
Table (5-8):	Measured Cross Sections Values for Production of ^{95m}Tc, and Isomeric Ratios on Natural Mo Targets at Different Energies.	149
Table (5-9):	Measured Cross Sections Values for Production of $^{96m+g}$Tc, on Natural Mo Targets at Different Energies.	151
Table (5-10):	Measured Cross Sections Values for Production of ^{99m}Tc on Natural Mo Targets at Different Energies.	154
Table (5-11):	Measured Cross Sections Values for Production of ^{152g}Tb, on Natural Gd Targets at Different Energies.	156
Table (5-12):	Measured Cross Sections Values for Production of ^{154g}Tb, on Natural Gd Targets at Different Energies.	158
Table (5-13):	Measured Cross Sections Values for Production of ^{154m}Tb, and Isomeric Ratios on Natural Gd Targets at Different Energies.	160
Table (5-14):	Measured Cross Sections Values for Production of ^{155}Tb, on Natural Gd Targets at Different Energies.	162
Table (5-15):	Measured Cross Sections Values for Production of ^{156}Tb, on Natural Gd Targets at Different Energies.	163
Table (5-16):	Measured Cross Sections Values for Production of ^{160}Tb, on Natural Gd Targets at Different Energies.	165

1-1 Introduction

Through this study many questions could be declared concerning the cross sections as a physical values, also the way in which to use them in practical applications or to assess their accuracy. All these questions can be answered by recourse to simple theoretical principles. Actually, the basic knowledge is needed even to ask a question in clear form about a cross section. Studying the nuclear reactions induced by charged particle is required to obtain data about the probability of emission of ejected particles or nuclei. As well as the concern to produce residual nucleus that involves the population of discrete, or discrete collective or collective levels in the compound nucleus that leads to the formation of radionuclide.

Medical applications of nuclear radiations are of considerable interest for the humankind nowadays. Cyclotrons and accelerators, available in recent years in an increasing number all over the world, are being used for the production of radioisotopes for both diagnostic and therapeutic purposes. The physical basis of this production is described through interactions of charged particles, such as protons, deuterons and alpha particles, with matter. These processes have to be well understood in order to produce radioisotopes in an efficient and clean manner. In addition to medical radioisotope production, reactions with low energy charged particles are of primary importance for two major applications. Techniques of ion beam analysis use many specific reactions to identify material properties, and thin layer activation technique as a probe for wears and corrosion assessment.

Thus it became clear that there are definite needs for a clear and concise explanation of the principles of cross section theory, accurate measurements, and reliable compilations of data. This is the first objective of this work on an experimental base.

The second objective of this work also is to present the principles of cross section measurement and use, as well as sufficient theory prediction so that the general behavior of cross sections is made understandable.

In recent years, low energy small cyclotrons have been installed in several countries. Producing standard gamma emitting radioisotopes such as ^{67}Ga, ^{111}In, ^{201}Tl and ^{123}I commonly employed in medical diagnostic investigations using gamma cameras "Single Photon Emission Computed Tomography" (SPECT). Although the production methods are well established, there are no evaluated and recommended nuclear data sets available for production routes of many radionuclides.

The need for standardization of methods, techniques, data evaluation, data analysis have become necessary [cf. 1, 2]. Many of important radionuclides production routs are still under investigation. Others need to improve the precision of their existing data or to extend the covered energy ranges. For quantitative study of nuclear reactions, the cross section values for both ejected particle and residual intimate nucleus at certain incident energy have been considered. In this chapter, a review for reaction cross section explanation and the necessary formulae are given.

1-2 Reaction Q-Values, and Thresholds

During radionuclide production with accelerators the reaction depends on energy threshold value, cross section which depends in its turn on the energy of the incident projectile. Usually at these accelerators light charged particles (proton, deuteron, ^{3}He, or α) are used as bombarding particles. A nuclear reaction A(x,y)B is generally described in terms of the cross section. This quantity essentially gives a measure of the probability for a reaction to occur and may be calculated.

In a nuclear reaction:

$$x + {}^{A}_{Z}A \rightarrow {}^{A'}_{Z'}B + y + Q \qquad (1-1)$$

The number of nucleons, charge, energy, momentum, angular momentum, parity, and isospin are conserved. The conservation of energy can describe the Q-value of a reaction. It can be derived from the difference in the nucleic masses and is an indication whether the reaction is energetically possible or not. However a minimum energy is required to initiate a reaction and called the threshold energy E_{th}, can be described by:

$$E_{th} = |Q| \cdot \left(1 + {m_x}/{M_A}\right) \qquad (1-2)$$

where m_x and M_A are the masses of the projectile and target nucleus, respectively. The energy liberated in the reaction is defined by the quantity Q, its value is given by:

$$Q = \left[E_B + E_y - E_x\right] \qquad (1-3)$$

Or,

$$Q = \left[(m_x + M_A) - (m_y + M_B)\right] \cdot c^2 \qquad (1-4)$$

where E_x, E_A, E_y, E_B refer to the kinetic energy of the projectile x, the target A, the products y, and B, respectively. m_xc^2, M_Ac^2 are the rest mass energy of the projectile and the target, and m_yc^2, M_Bc^2 are the rest mass energy of the products.

It should be pointed out that most of the experimental studies are carried out in the laboratory frame of reference (Laboratory System, LS) in which the target is at rest. Theoretical studies are more suitable in a center of mass reference frame (Center of Mass System, CMS) in which one center of mass represents the reactants and the other corresponds to the products. In most of the nuclear reactions the mass of the projectile is usually smaller than that of the target nucleus, so that for low incident energy the center of mass almost coincides almost with the coordinate of the target nucleus, i.e., LS and CMS practically coincide. The difference between these two frames becomes large for high-energy reactions, when the kinetic energy of the projectile exceeds the sum of rest masses of both colliding particles. In this case, CMS frame moves relative to LS frame with a velocity close to the velocity of light.

If $Q > 0$, the reaction is termed exoergic, i.e. accompanied by liberation of energy. If $Q < 0$, absorption of energy is needed to start the reaction; it is termed endoergic. For elastic collision, $Q = 0$. Exoergic reaction and elastic scattering can take place at any energy of the incident particle. For an endoergic reaction ($Q < 0$) the bombarding particle should have an energy value sustaining the absorbed energy due to atomic mass difference, this is characterized by the threshold energy. Note that the threshold energy, E_{th}, generally does not coincide with the reaction energy $|Q|$.

For the reactions in which the projectile mass is considerably less than the mass M_A of the target nucleus, the threshold practically coincides with $|Q|$. From Eq. (1-2) at $m_x \ll M_A$, this leads to E_{th} ? $|Q|$.

Due to Coulomb potential between the projectile and the target nucleus, additional energy should be added to the projectile energy, namely E_c. This energy depends on the charges of the projectile (Z_xe) and the target nucleus (Z_A e). where Z_x, Z_A, and e are the atomic number of the projectile, the atomic number of the target, and the electron charge, respectively, and the minimum distance of impact (r_{min}):

$$E_c = k \cdot \frac{Z_x \cdot Z_A \cdot e^2}{r_{min}} \qquad (1-5)$$

where $k = \frac{1}{4\pi\varepsilon_o} = Coulomb\,Constant = 8.988\times10^9\ N\cdot m^2/C^2$, ε_o is the permitivity of the free space $r_{min} \simeq R_x + R_A$, R_x and R_A are the radii of the colliding particle and the target nucleus, respectively. By using the electromagnetic radius formula:

$$R = r_0 \cdot A^{1/3} \qquad (1-6)$$

where $r_o \simeq 1.2\times10^{-13}\ cm$ and A is the mass number, finally we get:

$$E_c = k\cdot\frac{Z_x\cdot Z_A\cdot e^2}{r_0\cdot\left(A_x^{1/3}+A_A^{1/3}\right)} \qquad (1-7)$$

The energy thresholds for proton p on target T and Incoming particle Coulomb Barrier (ICB) are calculated with kinematics in the laboratory system [cf. 3]. By the mass defects from Firestone [cf. 4], and Eqs (1-2), and (1-8) in which the charges Z are non dimensional, nuclear radii are approximated as $\approx 1.4\ A^{1/3}\times10^{-15}$ m, and $e^2 = 1.44\times10^{-15}$ m MeV (let us remember that $1fm=10^{-15}$ m and $1b=10^{-28}\ m^2$):

$$ICB \approx \left(1.44/1.4\right)\left[\left(M_x+M_A\right)/M_A\right]\times\left[Z_xZ_A/\left(A_x^{1/3}+A_A^{1/3}\right)\right] \quad (MeV) \qquad (1-8)$$

In case of (p,xn) reactions, the ICB values varies in the range 7.92 MeV for (^{92}Mo) to 7.73 MeV for (^{100}Mo), thus their values are larger than some (p,n) reaction thresholds. Nevertheless as it happens in most similar cases, several experimental data show that below the (p,n) reaction barrier cross-sections are not negligible, due to the tunneling phenomena. In the case of production of metastable nuclide production, the energy of metastable level must be added to calculated Q-value. Finally, in the case of cluster emission, the Coulomb Barrier of the outgoing particle (OCB) must be added to the calculated E_{th} values. The OCB value is calculated by an equation similar to Eq. (1-8), in which, both masses, charges, and mass numbers of reaction products must be used, and is given by:

$$ICB \approx \left(1.44/1.4\right)\left[\left(M_y+M_B\right)/M_B\right]\times\left[Z_yZ_B/\left(A_y^{1/3}+A_B^{1/3}\right)\right] \quad (MeV) \qquad (1-9)$$

1-3 Status of Available Nuclear Data

All of the radioisotopes used for in-vivo studies can be arbitrary divided into five groups according to their chemical behavior, and biological function or mode of formation.

(1) The “Organic” short-lived β^+ emitters are ideally suited for labeling biomolecules and find their applications in PET (Positron Emission Tomography).

(2) Radiohalogens may also be regarded as “organic” isotopes since they are also useful for labeling biomolecules; some of them are suitable for in-vivo studies using the conventional γ-Cameras while the others find their applications in PET.

(3) Rare-gases are generally used for ventilation studies since many of radiohalogens are formed via rare-gas precursors, these two types of radionuclides are grouped together. Short-lived generator radionuclides are practical for general medicine use since an on-site or nearby cyclotron is not required.

(4) Alkali and alkali-like metals find their applications in myocardial perfusion studies.

(5) The List of “Inorganic” Radionuclides is large but their applications are limited.

Table (1-1) gives a list of the classified in-vivo radionuclides according to their chemical behavior [cf. 5]. Each radionuclide can be generally produced via several nuclear reactions. Not every reaction is suitable for large-scale production of a particular nuclide. Apart from cross section data others considerations must be available as ease of target construction, capability of withstanding high beam currents, as well as the subsequent chemical processing.

Table (1-2) shows the potential of a small four particle cyclotron with regard to medical radionuclides production. In practice one of the particles (p, d, ^{3}He, ^{4}He) is used in the medium and heavy mass regions, the proton mostly gives a high yield. In recent years ^{3}He-particle have found increasing applications since the energies of $^3He^{2+}$-particles available at compact cyclotron are generally higher than those of other particles.

Table (1-1): List of the Classified In-Vivo Radionuclides According to Their Chemical Behavior [cf. 5].

Types of Radionuclides for In-Vivo Studies
1. "Organic" Short-Lived β^+-Emitters: ^{11}C, ^{13}N, ^{15}O, ^{18}F, ^{30}P.
2. Halogens and Rare Gases: ^{18}F, ^{34m}Cl, $^{75,77}Br$, ^{125}Xe.
3. Generator Radionuclides: [produced by cyclotrons] (Diagonostic i.e. SPECT). $^{195m}Hg/^{195m}Au$, $^{109}Cd/^{109m}Ag$, $^{178}W/^{178}Ta$. (Therapeutic radionuclide i.e. β^--emitters) ^{212}Pb-^{212}Bi, (PET) they are β^+-emitters generators. $^{44}Ti/^{44}Sc$, $^{52}Fe/^{52m}Mn$^, $^{62}Zn/^{62}Cu$^, $^{68}Ge/^{68}Ga$*, $^{72}Se/^{72}As$, $^{82}Sr/^{82}Rb$*, $^{128}Ba/^{128}Cs$.
4. Alkali and Alkali Like Metals: $^{38,43}K$, ^{81}Rb, $^{128,129}Cs$, ^{201}Tl.
5. "Inorganic" radionuclides: ^{28}Mg, ^{45}Ti, ^{48}Cr, ^{67}Ga, ^{111}In, ^{73}Se, ^{97}Ru.

*Commercially available; ^Have been used clinically

Table (1-2): Potential of a Small Four Particles Cyclotron with Regard to Medical Radionuclides production [cf. 5].
(E_p=18MeV;E_d=10MeV;$E_{^3He}$=27MeV, E_α=20MeV)

Radio-nuclide	$T_{1/2}$	Nuclear Process	Radionuclide	$T_{1/2}$	Nuclear Process
Proton induced reactions			Deuteron induced reactions		
^{11}C	20m	$^{14}N(p,\alpha)$	^{15}O	2m	$^{14}N(d,n)$
^{13}N	10m	$^{16}O(p, \alpha)$	^{18}F	110m	$^{20}Ne(d, \alpha)$
^{18}F	110m	$^{18}O(p,n)$*	^{67}Ga	78h	$^{66}Zn(d,n)$*
^{22}Na	2.6y	$^{22}Ne(p,n)$*	^{75}Br	1.6h	$^{74}Se(d,n)$*
^{67}Ga	78h	$^{67}Zn(p,n)$*	^{123}I	13.2h	$^{122}Te(d,n)$*
^{75}Se	120d	$^{75}As(p,n)$	^{3}He-particles induced reactions		
^{86}Y	14.7h	$^{86}Sr(p,n)$*	^{18}F	110m	$^{16}O(^3He,p)$
^{94m}Tc	52m	$^{94}Mo(p,n)$*	^{76}Br	16h	$^{75}As(^3He,2n)$
^{111}In	2.8d	$^{111}Cd(p,n)$*	^{77}Kr	1.2h	$^{76}Se(^3He, 2n)$*
^{123}I	13.2h	$^{123}Te(p,n)$*	^{97}Ru	2.9d	$^{96}Mo(^3He,2n)$*
^{124}I	4.2h	$^{124}Te(p,n)$*	α-particles induced reactions		
^{201}Tl	3d	$^{201}Hg(p,n)$*	^{30}P	2.5m	$^{27}Al(\alpha,n)$
			^{38}K	7.5m	$^{35}Cl(\alpha, n)$

* Using isotopically enriched target material.

1-4 Nuclear Data Needs, and Compilation

The term "*nuclear data*" is very broad, it includes any type of data originating from "*the decay of radioactive nuclei*" or from "*the interaction of nuclei with matter*". However, as shown in Table (1-3) all data can be generally grouped into two categories:

i) Nuclear Structure and Decay Data

ii) Nuclear Reaction Cross Section Data

From the viewpoint of nuclear physics and nuclear chemistry the database in nuclear structure, decay data, and nuclear cross section data considered to be sufficiently extensive. From the user-oriented point of view the existing nuclear structure and decay data are also extensive. Only in some special cases the branching ratios or gamma-ray transition intensities are less certain. On the other hand, the application-oriented nuclear reaction data are concerned, the available information is extensive in some cases and scanty in the others.

In the case of neutron induced reactions which uses one of the following four (n,γ), (n,p), (n,α) and (n,f) reactions, the cross-sections for all pertinent reactions concerned are well known. On the other hand, in the case of charged particle induced reactions, the integral cross-section data needed for short-lived radionuclide production using cyclotrons were initiated recently, and have received comparatively less attention so it needs exhaustive compilation and also a salient treatment.

During the past thirty years, many laboratories have reported a large body of experimental data relevant to medical radioisotope production, and the charged particle data centers have compiled most of these data. However, no systematic efforts have been devoted to their standardization. Such a task would be too ambitious for any single national laboratory, implying a need for well-focused international effort. Under these circumstances, the International Atomic Energy Agency (IAEA) decided to undertake and organize the Coordinated Research Project (CRP) on Development of Reference Charged Particle Cross-Section Database for Medical Radioisotope Production. The project was initiated in (1995).

Table (1-3): Types of Nuclear Data, and Their Main Applications [cf. 6].

Nuclear Data	
Nuclear Structure and Decay Data (Main application in choice of the radionuclides for medical use)	Nuclear Reaction Cross Section Data (Main application in production of the radionuclides)
Nuclear Levels Spins and Parities Decay Modes and Half-Lives Decay Energies (Q-values) Branching Ratios α, β, and γ Ray Energies Multipolarities of γ-Transitions Conversion Coefficients Fluorescence Yields X-rays Bremsstrahlung and Auger e^- Decay Schemes	Scattering and Capture Cross-Sections Resonance Integrals Fission Yields Range-Energy Relationship Q-Values Reaction Thresholds Angular Energy Distribution Data Differential and Integral Cross-Sections Excitation Function Thin and Thick Target Yields

In recent years the IAEA has been paying a considerable attention to the subject. It focused on radioisotopes for diagnostic purposes and on the related beam monitors reaction in order to meet the current needs. The CRP of the IAEA constituted the first major international effort dedicated to standardization of nuclear data for radioisotope production, and covered the following areas:

(1) Compilation of data on the most important reactions for monitoring light ion charged particle beams (p, d, ^{3}He, and α).

(2) Evaluation of the available data (both by fitting and theory).

(3) Compilation of production cross-section data on radioisotopes most commonly used in medicine.

(4) Development of calculation tools for predicting unknown data.

The Coordinated Research Program (CRP) of the IAEA recommended reaction cross sections for the production of medically important radionuclides have been made available in numerical form. The CRP produced a much-needed database and a handbook, covering reactions used for monitoring beam currents and for routine production of medically important radioisotopes. It is believed that the recommended cross-sections are accurate enough to meet the demands of all current applications, However, further development of evaluation methodology and more experiments will be needed for exact determination of the errors and their correlations related to beam monitor, beam energy, and detection efficiency. In addition to the reaction cross-sections data, yields of the produced radioisotopes could be calculated from the recommended cross-sections data.

Considering nuclear data involved in radionuclide production for different uses, in areas of interest, existing data from the evaluated files, recent measurements, and model calculations were examined. It was found that there is a clear need of precise nuclear data for many technologically important materials in the energy region where few nuclear data exist. Systematic calculations and predictions resulting from theoretical models may complement the data for the unmeasured energy regions and nuclei.

The systematic calculations are less accurate, however it is less laborious than the theoretical models method of predicting an unknown cross section. Nevertheless, such systematic trends allow a quick prediction of unknown cross section. In cases where the experimental measurements are extremely tedious or where theory cannot be applied with certainty, the systematic calculations, if done with precision and caution, could provide useful information on the various competing reactions.

1-5 Cyclotron Production of Medically Used Radionuclides

The use of radionuclides as tracers in medicine (nuclear medicine) is a well established field of science, with a long history [cf. 7]. The spatial distributions of labeled compounds in the human body are of much importance. The comparatively recent developments in instrumentation for imagining have shifted the forefront of the research away from simple qualitative distribution studies towards quantifying biochemical pathways in the human body. Positron Emission Tomography (PET), and Single Photon Emission Computed Tomography (SPECT) are the most common techniques have used in nuclear medicine.

The major use of radionuclides in medical applications is in-vivo imaging. The effective use of PET or SPECT scanners is dependent on the availability of suitable radioactively labeled compounds of sufficient quality (chemical, radiochemical, and pharmaceutical), and quantity. For PET applications, the positron emitting radionuclides that are used as precursors for radiotracer synthesis are readily produced with a charged particle accelerator through nuclear reactions during bombardment of a target material. Produced radionuclides are used either immediately after the end of bombardment, or even on-line in some special cases. Also generator systems can be used, where a longer lived “parent” radionuclide decays to a shorter-lived “daughter” radionuclide. For SPECT applications, single photon emitting radionuclides are produced both with particle accelerators and through neutron bombardment and fission in a nuclear reactor.

The PET or SPECT facility requires: an accelerator (small cyclotron), a chemistry laboratory for radiotracer and/or radiopharmaceutical synthesis and a radiography camera, PET Camera for Positron Emission Tomographs (PET) or Gamma Camera for Single Photon Emission Computed Tomographs (SPECT), see Figure (1-4). An integral part of a cyclotron facility is a team of specialist consisting of experts in biochemistry, chemistry, computer science, instrumentation, mathematics, medicine and physics.

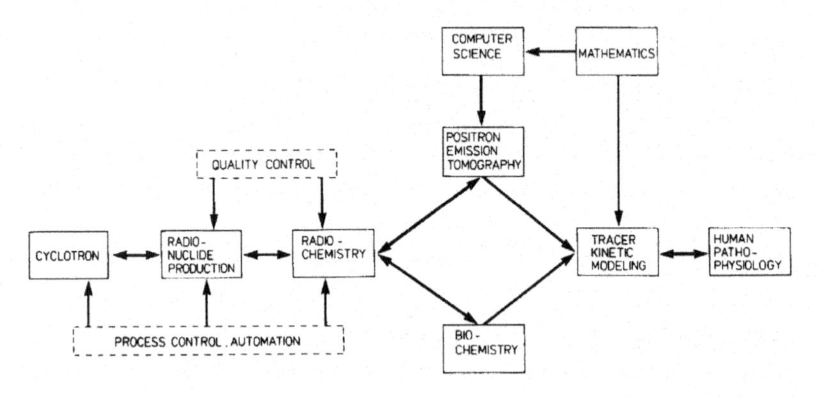

Figure (1-1): Scheme of Disciplines and Activities Involved in a Cyclotron Facility.

The choice of a suitable radionuclide for certain medical application depends on several nuclear and chemical aspects. The nuclear decay data of radionuclide help to decide whether they can be used for therapeutic or diagnostic purposes. The medical requirements impose many restrictions on the half-life of the tracer, its biological half-life, its path to the targeted organ, the way of its absorption and distribution in the organ, the energy of its radiation, the requires specific activity, etc. On the other hand, the chemical characteristics of the radionuclides are very important to be used in their proper function. Important parameters of radionuclide production are:

(1) Yield of the desired radioisotope

(2) Chemical form

(3) Chemical purity

(4) Radiochemical purity

(5) Radionuclidic purity,

(6) Specific radioactivity

As all important positron emitting radionuclides are short lived ($T_{1/2}$<3h), the requirements on the synthetic chemistry for the labeled tracers are rather special. Fast chemical reactions and methods are therefore required. It is mandatory that the production of the short-lived isotopes is closely interrelated to the radiochemical work. Some important radionuclides used for PET imaging are listed in Table (1-4).

Table (1-4): Presents a list of some Important Positron Emitting Radionuclides Used for PET, their Half Lives and Typical Nuclear Reactions for Their Production are given [cf. 5].

Nuclide	$T_{1/2}$ [min]	Nuclear reaction (typical)	Type of cyclotron used in production (see next section)
^{11}C	20.4	$^{14}N(p,\alpha)$ or $^{11}B(p,n)$	Level I or II
^{13}N	10.0	$^{16}O(P,\alpha)$ or $^{13}C(p,n)$	Level I or II
^{15}O	2.1	$^{14}N(d,n)$ or $^{13}N(p,n)$	Level I or II
^{18}F	19.8	$^{20}Ne(d,\alpha)$ or $^{18}O(p,n)$	Level I or II
^{38}K	7.6	$^{40}Ca(d,\alpha)$ or $^{38}Ar(p,n)$	Level II
^{68}Ge*/^{68}Ga	68.3	$^{69}Ga(p,2n)68Ge$	Level III
^{75}Br	98	$^{76}Se(p,2n)$ or $^{75}As(^{3}He,3n)$	Level III
^{82m}Rb*/^{82}Rb	1.3	$Mo(p,spall)^{82}Sr$ $Rb(p,xn)^{82}Sr$	Level IV

* Marks refer to radionuclide generator product

Among the entire important positron emitters only ^{18}F can be produced in quantities with sufficiently high specific radioactivity for PET [cf. **8**], using a nuclear reactor. Other positron emitters must be produced with particle accelerators, preferably a cyclotron. The recent development in radionuclide production for PET has been directed towards small in-hospital cyclotrons [cf. **9**].

Cyclotrons used for radionuclide production are usually classified in four levels, I-IV, depending on their maximum proton beam energy [cf. **5**,**8**]. The higher levels III and IV cyclotrons can mostly be found in big research centers or big industrial laboratories. However, for the production of short-lived positron emitter radionuclides only levels I and II cyclotrons are suitable, see Table (1-5).

Some small cyclotrons are exclusively devoted to the production and on-site use of short-lived β^+-emitters. Medium and larger-size machines have often-interdisciplinary utilization, with a considerable fraction of beam-time for radionuclide production, they are capable of delivering a variety of radionuclides.

Table (1-5): Presents a list of some Important PET, and SPECT Radionuclides, and a Classification of Cyclotrons Used in Their Production [cf. **5**].

Classification	Characteristics	Energy (MeV)	Major Radionuclides Produced
Level I	Single Particle, (p, d)	$E_{p,d} \leq 10$ (MeV)	^{11}C, ^{13}N, ^{15}O, ^{18}F
Level II	Single or Multiple Particles (p, d, 3He, α)	$E_p \leq 20$ (MeV)	^{11}C, ^{13}N, ^{15}O, ^{18}F, ^{67}Ga, ^{75}Se, ^{111}In, ^{123}I, ^{124}I, ^{201}Tl,
Level III	Single or Multiple Particles (p, d, 3He, α)	$E_p \leq 40$ (MeV)	^{38}K, ^{73}Se, $^{75-77}Br$, ^{123}I, ^{67}Ga, ^{111}In, ^{201}Tl, ^{22}Na, ^{57}Co, $^{81}Rb(^{81}Kr)$
Level IV	Single or Multiple Particles (p, d, 3He, α)	$E_p \leq 100$ (MeV)	^{123}I, $^{72}Se(^{72}As)$, $^{82}Sr(^{82}Rb)$, ^{117m}Sn
Level V	Single or Multiple Particles (p, d, 3He, α)	$E_p \geq 200$ (MeV)	^{26}Al, ^{32}Si, ^{44}Ti, ^{67}Cu, $^{68}Ge(^{68}Ga)$, ^{95m}Tc $^{82}Sr(^{82}Rb)$, ^{109}Cd,

The development of the Anger camera [cf. **10**] and the ^{99m}Tc-nuclide available from the ^{99}Mo generator [cf. **11**-**13**] for the labeling of compounds further strengthened the position of the single photon emitting radionuclides in the field of nuclear medicine [cf. **14**]. The development of coincidence techniques for the measurement of annihilation radiation produced in positron emission, and the fact that the radionuclides of carbon, nitrogen and oxygen emit positrons, gave the impetus for a renewed interest in accelerator produced radionuclides in the early seventies [cf. **15**]. Some special advantages of the cyclotron products as compared to reactor products are:

(1) Possibility of in-vivo quantitative study of regional physiological functions using organic radionuclides β^+-emitters,
(2) High specific activity of the produced radionuclide,
(3) Low radiation dose to patient.

1-5.1 Target Chemistry, and Targetry

Target chemistry is controlling the form of the radioisotope after the bombardment is over. The chemical form of the element is chosen to have good mechanical and thermal properties. The irradiated material could be solid, liquid or gaseous. During irradiation, the target is exposed for a long period to high-intensity particle beam, neutron and gamma ray fluxes. Targetry is meant by the thermal stability of the target and its contents, these include the window material, the interior surface of the target holder, and the material used for construction. The radiolytic and chemical stability of the contents is a function of both the state (gas, liquid, and solid), and reactivity of the contents.

In order to put this complex subject into perspective, practical yields of radionuclides in elemental or simple compound form is a function of the following parameters [cf. **14**, **16**, **17**]:

(1) Energy of the beam impinging on the target, high melting point, and high thermal conductivity
(2) Target optimum current,

As a function of chemical processes could take place in target that imply no decomposition on heating, no gaseous byproducts, and no chemical reaction with the encapsulating material.

As a function of target design (cooling, etc.)

(3) Time of irradiation

Optimization of yield versus on-time of the cyclotron,

Optimization of yield because of dynamics of reactions in target versus time.

(4) Activity recoverable from target
(5) Chemical and physical form required, that will permit simple chemical and physical manipulation to remove the target from the irradiation capsule, and which will subsequently allow post irradiation processing to be performed simply and rapidly with remote handling
(6) Dependence in turn on possible on-stream chemical conversion.
(7) Stable to irradiation environment

A successful radionuclide production is crucially dependent on the quality and composition of the target materials. The radionuclidic composition of the product is of course influenced by the isotopic composition of the target itself. By using isotopically enriched (depleted) targets both the yield and the radionuclidic purity of a given product can be increased. By the use of additives to the target the chemical form of the product can be influenced. Minute amounts of chemical contaminants in the target can have a detrimental effect both on the specific radioactivity and chemical composition of the product.

1-5.2 High Current Irradiations at Cyclotrons

The target holder is used mainly to support the irradiated material, activity handling, and prevents overheating or evaporation which is achieved using efficient cooling system, and suitable design ensure quick transportation. The target holder should satisfy the following conditions:

- Its material producing low background radioactivity
- Simple loading and unloading mechanisms of the irradiated target from its holder
- Good thermal contact between the holder and the irradiated material
- Electrical isolation from the surroundings for beams current monitoring.

In routine production the design of the cyclotron target is more complicated than in the test experiments. Pneumatic arms and remote control are usually used for loading and unloading of the target. For example, if the target material does not have good thermal properties, the cooling is applied on both sides of the target material [cf. **15**], which makes the design more complicated.

To decrease the heat effect of the beam on the target material several laboratories used inclined targets relative to the beam direction [cf. **18**]. In inclined position the path of the projectile within the target material becomes longer than in the normal position. This allows use of a thin layer of the target material. In addition, the power deposited per unit area is less than that for normal position, and the cooling becomes more efficient.

The absorbed energy within the target produces thermal power per elemetary charge which, can be estimated from the following formula:

$$Power\,(Watt) = \frac{Absorbed\ Energy\,(MeV) \cdot Beam\ Current\,(\mu A)}{Elementary\ Charge\,(e)} \qquad (1-10)$$

The desired energy range for production determines the thickness of the target. It is calculated from the tabulated stopping power or the range values. Heat transfer in solids is somewhat simpler than in other media since the heat usually flows through the target matrix mainly by conduction. The heat will be conducted from the hotter region of a material to the cooler region according to Fourier's Law, which is in one dimension.

$$Q_{cond} = -k \cdot A \cdot dT/dx\Big|_{T_1}^{T_2} \qquad (1-11)$$

where,
Q_{cond} = heat transferred by conduction in (watts)
A = cross section area (cm^2)
k = coefficient of thermal conductivity (watts/cm·K)
dT = temperature difference in Kelvin; dx = differential distance (cm)
If this equation is integrated holding k, A, and Q constants, then the result is the heat transfer equation in one dimension given by:

$$Q_{cond} = -k \cdot A \cdot (T_1 - T_2)/x \qquad (1-12)$$

T_1 = temperature of the hotter part (K);
T_2 = temperature of the cooler part (K).

Once the heat has been transferred to the cooled surface of the target, the heat will usually be removed by a fluid such as water flowing against the back of the target. For a compound target the stopping power is estimated for every elemental composition over the desired energy range. The resulting values are added according to the fractional composition of every element in the target.

In some instances, the thermal properties of the compounds or alloys may be more favourable than that of their respective pure elements. For example, many metals of low melting temperature form oxides whose melting points are very high. As an example, tellurium metal melts at about

450°C, while the melting point of its oxide, TeO_2, is 733°C and its heat conduction properties is very poor.

1-6 Calculation of the Activation Cross Sections

The reaction cross section refers to the probability that the reaction of interest can occur. The cross section for the reaction A(x,y)B is denoted by σ and is defined as the number of reactions (x,y) for one target nucleus placed in unit incident flux. The cross section is measured in barns ($1b=10^{-28}$ m^2). It does not represent the geometrical effective area over which the reaction occurs.

The reaction cross section is a string function of the incident particle energy to induce a certain reaction. It is known as an integral cross section when integrated over all angles. The variation of the cross section with incident energy is called the excitation function. The reaction cross section is calculated for a certain type of the produced nuclide, and depend specifically on the final state of this nuclide. So, it is namely called reaction channel for specific initial conditions (i.e. energy, projectile type, polarity, etc.,) of a nuclear reaction, determining the cross section for that channel, differs whenever gives a nuclide in ground or metastable states, where there are different competing processes could take place. Accompany a collision between a proton beam of considerable energy with target nucleus variety of channels could be open and relieve cross section for every channel simultaneously. If the energy of the projectile is less than the threshold of some endoergic reaction, then it is called a closed channel. At very low energies of the projectile, only the elastic and exoergic channels are open. As the energy increases, the endoergic reaction channels become energetically possible, one after the other, according to their respective energy thresholds.

The excitation functions for nuclear reactions are utilized when designing the strategy for a production route. Important considerations are particle energies available, the yield of the desired product and the suppression of unwanted side reactions [cf. **19**]. Two standard methods commonly used for charged particle induced reaction cross section measurements. The first one is applicable in the case of variable energy cyclotrons. Thin samples of the target material are irradiated either with extracted beams of varying energies or at appropriate radius in the internal beam. At external beam thin foil enable direct detection of the emitted particles and identification of the activated products. The direct measurement is also known as the spectral measurement technique, these methods have their own advantages and limitations. An outline of the characteristics of the two methods is given in Table (1-6). The second method is known as the "stacked-foil" or "stacked-pellet" technique and is more commonly used with the direct detection of residual activity. A set of foils or pellets in a stack is irradiated whereby the incident particle exhibits fractional loss of its energy at each foil. Therefore it is possible to produce activity at different energy values over a certain energy range. Plotting of the calculated cross section deduced from the absolute activity, as a function of incident energy of the projectile is known as excitation function.

Both techniques are commonly used for determination of the excitation function. The number of interactions N′ taking place per unit time when N nuclei are bombarded by projectiles of density n per cm^3 and velocity v, cm per sec, is given by [cf. **20**]:

$$N' = N \cdot n \cdot v \cdot \sigma, \qquad (1-13)$$

Assuming that N′ is small enough so that the nuclei do not change the projectile density appreciably, thus the interaction rate is proportional to the product of projectile density n and velocity v, which is the flux density I. For simplicity it may be called projectile flux. Thus a unit cross section σ is one that gives one interaction per unit time when a unit projectile density is incident on a single nucleus.

It is noted that the interaction rate depends on the total number of nuclei, N, not on the number per cm^3. From this point the simple concept of reaction cross-section as nuclear area could be declared as following. The number of projectiles in the volume vσ will hit a given nucleus per second. This number is nvσ for one nucleus or Nnvσ for all nuclei, regardless of the density of the sample or orientation, if N is sufficiently small so that there is no self-shielding. Although the practical utility of the concept of a cross section as nuclear area is obvious, its unreality is clearly shown by the variation in size of cross sections, enormous compared to the relatively constant true nuclear size. From the measured radioactivity the cross section is obtained using the well known activation formula [cf. **20**]:

$$A = I \cdot N \cdot \sigma \cdot \left(1 - e^{-\lambda \cdot t}\right) \qquad (1-14)$$

where
A = the activity of the produced radionuclide (Ci),

I = flux intensity of the incident particles (number of particles/cm^2. sec),

N = number of target i.e. surface density (atoms/cm^2),

σ = cross section of the reaction (cm^2),

λ = decay constant of the resulting nuclei (sec^{-1}),

t = irradiation time (sec).

Table (1-6): Characteristics of the Direct Detection of the Emitted Particles, and Activation Technique.

CHARACTERISTICS	EXPERIMENTAL TECHNIQUES	
	Direct Detection of the Emitted Particle	Activation Technique
Type	On-Line	Off-Line
Detection Selectivity	Particle	Product
Target	Must be thin	Can be thick
Product	May be stable or radioactive	Has to be radioactive
Measured Data	Double differential data	Integral data

In order to calculate the cross section of a proton-induced reaction, the well-known activation formula is considered. In the basic underlying equation the production rate of a radioactive nuclide is described as a function of the irradiation time [cf. **21**]:

$$\frac{dN_i}{dt} = N_t \, \sigma_i \, \varphi(t) - \lambda \, N_i(t) \qquad (1-15)$$

where N_i, is the number of induced atoms (produced), N_t the number of target atoms, σ_i the reaction cross section, λ_i the decay constant and $\varphi(t)$ the time dependent particle flux. The increase in N_i due to the activation process is expressed as the first term whereas the second term denotes the decrease resulting from the decay of N_i.

To obtain the number of reaction products N_i at time t, the above equation is solved with the initial condition: $N_i = 0$, and $\varphi = \varphi_0$ at $t = 0$ and t_i being the activation time (time of irradiation), and is expressed as the following:

$$N_i(t) = N_t \sigma_i \varphi_o \int_0^{t_i} \exp(-\lambda_i t)\, dt \qquad (1-16)$$

At the time t_d (decay time) which is the time elapsed from the end of activation time and beginning of counting, the activity is give by:

$$N_i(t) = \frac{N_t \sigma_i \varphi_0}{\lambda_i}\left[1 - \exp(-\lambda_i t_i)\right] \qquad (1-17)$$

$$\lambda_i N_i(t) = A_{irr} = N_t \sigma_i \varphi_0 \left[1 - \exp(-\lambda_i t_i)\right] \qquad (1-18)$$

where A_{irr} is the activity at the end of irradiation. The induced activity becomes A_d at the time t_d after the end of irradiation:

$$A_d = A_{irr} \exp(-\lambda_i t_d) \qquad (1-19)$$

The net counts N of a full energy peak of gamma-transition of the same nuclide for a counting time t_c is given by:

$$N = \int_0^{t_c} A_{irr} \exp(-\lambda_i t)\, \varepsilon_\gamma I_\gamma\, dt \qquad (1-20)$$

where ε_γ is the peak efficiency of the detector with necessary corrections, I_γ is the absolute branching ratio for a particular γ-energy of the produced nuclide. Integrating Eq. (1-20), the activity at the end of irradiation is:

$$A_{irr}=\lambda_i N_i(t_i,t_d)=\frac{\lambda_i N}{(1-\exp(-\lambda_i t_c))\varepsilon_\gamma I_\gamma} \qquad (1-21)$$

The induced activity at the time t_d after the end of irradiation becomes:

$$A_{irr}=\frac{\lambda_i N}{(1-\exp(-\lambda_i t_c))\varepsilon_\gamma I_\gamma}\cdot\exp(\lambda_i t_d) \qquad (1-22)$$

Combining Eq. (1-18), and Eq. (1-22), we obtain:

$$\sigma=\frac{\lambda_i N\exp(\lambda_i t_d)}{N_t\,\varphi\,\varepsilon_\gamma I_\gamma(1-\exp(-\lambda_i t_i))(1-\exp(-\lambda_i t_c))} \qquad (1-23)$$

Which is the required activation formula for the determination of cross section.

In principle, the activation technique allows a more precise determination of the cross sections when the residual nucleus is radioactive and emits easily measurable radiation [cf. **20**,**21**]. At proton energies near 18 MeV, several reaction channels are likely to be open. For example the (p,xn), cross section is the sum of (p,n), (p,2n), ...etc. The activation method does not allow a distinction between (p,d) and (p,np) processes, as all of these lead to the same final nucleus. Often one of the residual nuclei reached by proton or alpha particle decay channel is stable, making an activation measurement incomplete.

Another approach to the activation method involves the combination of radiochemical separations. In case physical methods of measurement alone are mostly insufficient; good radiochemical separations together with high selective counting methods are required to identify the radioactive products. The radiochemical technique is needed especially when:

i) Reactions with low yield are studied,

ii) Thin samples are prepared for beta and X-ray measurements to reduce self-absorption effects,

iii) Short-lived isotopes have to be separated from strong long-lived matrix activities (overlapping gamma-rays and high detector dead time.

The criteria for a good radiochemical separation are:

(1) Efficient removal of radioactive impurities,

(2) High chemical separation yield,

(3) Simplicity, reproducible, and fast chemistry.

1-6.1 Cumulative Cross-Sections

The equation (1-23) is strictly valid only for the so-called independently produced radionuclides because the only production mechanism is the nuclear reaction leading to the produced nuclide. But in the majority of cases a further production by β^-, β^+, EC, or α-decays of a radioactive precursor has to be taken into account. Since there are sometimes ambiguities existing about the terms independent and cumulative cross-sections we have to give some clarifications.

A cross-section for the production of a nuclide is denoted as independent if the nuclide can only be produced directly via the nuclear reaction between the projectile and the target nucleus and not via subsequent β^-, β^+, EC, or α decays. Such independent cross-sections are obtained if:

1. The nuclide is shielded by nuclides that are stable against β-decay,
2. The nuclide is shielded by a long-lived progenitor (e.g. ^{60}Co by ^{60}Fe with $T_{1/2} = 1.5 \times 10^6$ y, or ^{194}Au by ^{194}Hg with $T_{1/2} = 520$ y),
3. The cross-section for the production of a progenitor is also measured so that the production via decay can be corrected.

In all other cases the cross-sections are cumulative since they include also the production via decay of precursors. If we consider, e.g. the production of a nuclide D (daughter) on the one hand by the nuclear reaction and on the other by decay of a radioactive precursor M (mother). Then the solution of the differential equation corresponding to Eq. (1-24) for the activity $A_D(t)$ of D for times $t > t_{irr}$ is given by [cf. 3]:

$$A_D(t) = N_T\Phi\left(\begin{array}{l}\left(\sigma_D + \sigma_M \cdot \frac{\lambda_M}{\lambda_M - \lambda_D}\right)\left(1-\exp\left(-\lambda_D t_{irr}\right)\right)\exp\left(-\lambda_D t\right) \\ +\left(\sigma_M \cdot \frac{\lambda_M}{\lambda_M - \lambda_D}\right)\left(1-\exp\left(-\lambda_M t_{irr}\right)\right)\exp\left(-\lambda_M t\right)\end{array}\right) \quad (1-24)$$

where σ_D, σ_M are the independent cross sections for the daughter D and the mother M, respectively. Provided that the half-life of M is short compared to that one of D ($\lambda_M >> \lambda_D$), we can neglect the second term in Eq. (1-24) for large t_d. This yields:

$$\sigma_{D,cum} = \sigma_D + \sigma_M \cdot \frac{\lambda_M}{\lambda_M - \lambda_D} \quad (1-25)$$

With the cumulative cross-section $\sigma_{D,\,cum}$ of the nuclide D calculated according to Eq. (1-25).

The assumption of very short-lived progenitors holds in many cases, but not in all. There are cases {e.g. ^{86}Zr ($T_{1/2}$= 16.5 h) and ^{86}Y ($T_{1/2}$= 14.74 h), ^{88}Zr ($T_{1/2}$= 83.4 d) and ^{88}Y ($T_{1/2}$= 106.6 d), ^{95}Zr ($T_{1/2}$= 64.0 d) and ^{95g}Nb ($T_{1/2}$= 34.97d), ^{96m}Tc ($T_{1/2}$= 51.5 m), ^{96g}Tc ($T_{1/2}$= 4.28 d) and ^{96g}Tc ($T_{1/2}$= 4.28 d), ^{96g}Nb ($T_{1/2}$= 23.35 h)}. Assume a mother nuclide M of known activity A_M decaying with decay constant λ_M into the daughter D with λ_D for which wrong activities A^*_M are calculated according to Eq. (1-25), with a decay constant λ_M, comparable to λ_D. In this case use of Eq. (1-25) results in a wrong $A_p(t_{EOI})$ which here shall be denoted as $A^*_D(t_{EOI})$. The true $A_D(t_{EOI})$ is calculated from $A_D^*(t_{EOI})$ by Michel et al., (1997) [cf. **22**]:

$$A_D(t_{EoI}) = A_D^*(t_{EoI}) + A_M(t_{EoI}) \cdot \frac{\lambda_D}{\lambda_D - \lambda_M}$$

$$\times\left(1 - \frac{\lambda_D}{\lambda_M} \cdot \frac{1-\exp\left(-\lambda_M t_C\right)}{1-\exp\left(-\lambda_D t_C\right)} \cdot \exp\left(-\left(\lambda_M - \lambda_D\right)t_D\right)\right) \quad (1-26)$$

In this case the (cumulative) cross section of the mother nuclide can be measured and an independent cross section for the daughter product D is determined using the correction of Eq. (1-26). If the condition ($\lambda_M >> \lambda_D$) under which we derived the cumulative cross-section $\sigma_{D,\,cum}$ of the nuclide

D is fulfilled and we were able to measure σ_M, then we can derive the independent cross-section σ_D for the production of D from Eq. (1-26).

Higher orders of grandparent progenitors can be neglected in all cases because of the strong decrease of half-lives with increasing distance from the valley of stability.

Solving the system of coupled differential equations describing the decay of the mother and the decay and buildup of the daughter after the end of irradiation (t = 0) we calculate the corrected activity A_D according to Eq. (1-26) and obtain the independent cross-section of the daughter via Eq. (1-27) [cf. **22**]:

$$\sigma_D = \frac{A_D^*\left(t_{EoI}\right)}{N_T \Phi\left(1-\exp\left(-\lambda_D t_{irr}\right)\right)} - \sigma_M \times$$

$$\times\left(1-\frac{\lambda_D}{\lambda_D-\lambda_M}\left(1-\frac{1-\exp\left(-\lambda_M t_{irr}\right)}{1-\exp\left(-\lambda_D t_{irr}\right)}\right)\right) \qquad (1-27)$$

1-6.2 Isomeric Cross Section Ratios

When the produced nucleus is in an excited state, it is usually deexcited to the ground state in a time of the order of 10^{-14} seconds or less. In some cases however, the time may be long enough to be measured, ~ 0.1 seconds or longer. Such a nucleus is said to be in a metastable state and is distinguished from its ground state by an asterisk B*. The nucleus B of the same mass number and atomic number that differ in this way in their state of energy are called isomers. Moreover, the state itself is known as an isomeric state. The isomeric state decays either by gamma-ray emission to the ground state or by electron capture (EC), β or alpha particle emission to another nuclide. The process of reversion to the ground state by emission of the gamma ray is called the isomeric transition (IT).

For some nuclide, metastable states with long half-life can be formed and activation experiments can sometimes give information on the production of both the ground (σ^g) and metastable state (σ^m) cross sections. The cross sections for these individual discrete states are difficult to systematize since they are strongly dependent on the spin and parity of the states concerned [cf. 23-25]. Hence in contrast to the available systematic calculations of the integral cross sections of neutron induced reactions, those predicting the isomer cross section ratios for proton induced reactions are very scarce [cf. 24]. Therefore it is important to determine if there is any significant correlation of this ratio, if exists, to any parameter that could provide a model.

Systematic analyses of the isomeric cross-section ratios $\sigma^m/(\sigma^m+\sigma^g)$ for (n,2n) reactions were also performed using the pre-equilibrium and statistical Hauser-Feshbach code GNASH (Young 1977) [cf. 25] to study the dependence of the branching ratio on the isomeric spin by calculating the pertinent isomeric cross section ratio.

Recent mechanistic studies on isomer distribution in nuclear reactions have included investigations about the effect of:

i) Increasing projectile energy,
ii) Increasing mass and charge of the projectile,
iii) Role of different reaction channels leading to the same product nucleus,
iv) Low cross sections for the population of individual levels,
v) Necessity of X-ray spectroscopy due to the highly converted low energy transitions,
vi) Chemical separations and low level β^- counting due to the pure β^- emitters of some of the isomeric states.

1-7 Yield, Effective Cross Section

The term cross-section and yield, widely used in practical radioisotope production, often differ from basic definitions used in nuclear reactions theory. Different application-oriented groups use these terms in a non-standard way. In order to avoid misinterpretation of the data we briefly summarize definitions of the most important quantities describing nuclear reactions in the field of practical radioisotope production and activation technology.

1-7.1 Production Cross-Section

In isotope production and application of monitor reactions, usually the activity of the product radioisotope is measured. The related quantity of interest is then the integral cross-section or the production cross-section. It refers to a sum of cross-sections of all reaction channels on a well-defined target nucleus, which lead to direct production of the final nuclide. The same final nuclide can also be produced indirectly via the decay of progenitors produced simultaneously on the target nucleus. In many cases the separation of direct and indirect routes becomes not important and one uses the cumulative production cross-section to describe these two routes together. This becomes even more complicated when one uses a natural multi-isotopic target element where different reaction channels on different target nuclei can contribute to the production of the same final radioisotope. In this case one uses the elemental production cross-section to describe all production routes together.

It should be noted that in doing so one must properly calculate the number of target nuclei, by summing nuclei of all contributing target isotopes. If one considers also indirect production routes, the elemental cumulative production cross-section should be used.

Similarly, the notation isotopic production cross-section is used to describe reactions with mono-isotopic target elements. The present section aims to address the needs of everyday practice, where one uses elemental targets that are generally multi-isotopic and sometimes mono-isotopic. Throughout this section we consistently use the term cross-section. For beam monitor reactions, this term means cumulative elemental production or cumulative isotopic production cross-section of the final nuclide. For medical radioisotope production, this term means elemental production or isotopic production cross-section of the final nuclide.

1-7.2 Production Yield

A thin target has a thickness so small that the reaction cross-section can be considered as constant through the whole target. This is equivalent to the energy loss being negligible when compared to the energy range needed to see significant changes in the reaction cross-section. A thick target has its thickness comparable or larger than the range of the incident particle in the target material.

The yield for a target having any thickness can be defined as the ratio of the number of nuclei formed in the nuclear reaction to the number of particles incident on the target. It is termed as the physical yield, Y. It is customary to express the number of radioactive nuclei in terms of the activity, and the number of incident particles in terms of the charge. Thus, Y can be given as activity per Coulomb, in units of GBq/C. The analytical meaning of the physical yield is the slope (at the beginning of the irradiation) of the curve of the growing activity of the produced radionuclide versus irradiation time.

Radioisotopes disintegrate during the bombardment, therefore for practical applications other yield definitions are used taking into account this effect. The activity at the end of a bombardment performed at a constant 1 μA beam current on a target during 1 hour is closely related to the measured activity. In everyday isotope production by accelerators, the so called lμA-lh yield, A_1. In practice, this latter quantity can be used when the bombardment time is significantly shorter than or comparable with the half-life of the produced isotope.

When the irradiation time is much longer than the half-life of the produced isotope, a saturation of the number of the radioactive nuclei present in the target is reached, and their activity becomes practically independent of the bombardment time (at a constant beam current). This activity produced by a unit number of incident beam particles is the so-called saturation yield, A_2. There are close relationships between the above-mentioned yields. Using the decay constant of the radionuclide λ and the irradiation time t one gets:

$$Y = A_1 \frac{\lambda}{1-\exp(-\lambda t)} = A_2 \qquad (1-28)$$

Several other definitions are often used. Differential or thin target yield is defined for negligibly small (unit) energy loss of the incident beam in the thin target material. Thick target yield is defined for a fixed macroscopic energy loss, $E_{in}-E_{out}$, in a thick target. Integral yield is defined for a finite energy loss down to the threshold of the reaction, E_{in}-E_{th}. The thin target yield is easily related to the reaction cross-section and the stopping power of the target material for the beam considered [cf. **26**].

To this end, recommended cross-sections discussed in the present study were used. The Yield for any target thickness, Y_{thick} can be obtained from the simple formula:

$$Y_{thick}\left(E_{in}-E_{out}\right)=Y\left(E_{in}\right)-Y\left(E_{out}\right), \qquad (1-29)$$

where E_{in} is the incident particle energy and E_{out} is its outgoing energy. For a more detailed discussion and for practical calculations we refer to the extensive list of references in the literature [cf. **26**,**28**].

In addition, the target stopping powers of Ziegler [cf. **27**], and nuclear decay data of Firestone [cf. **4**] were used. The stopping-power of the target element could be calculated by the SRIM-2003 Code [cf. **27**]. SRIM-2003 is a software package concerning the stopping and range of ions in matter. Since its introduction in 1985, major upgrades are made about every five years.

For SRIM-2003, the following major improvements were made: (1) About 2200 new experimental stopping powers were added to the database, is increasing it to over 25,000 stopping values. (2) Improved corrections were made for the stopping of ions in compounds. (3) New heavy ion stopping calculations have led to significant improvements on SRIM stopping accuracy. (4) A self-contained SRIM module has been included to allow SRIM stopping and range values to be controlled and read by other software applications. A full catalog of stopping power plots has been published at www.SRIM.org. Over 500 plots show the accuracy of the stopping and ranges produced by SRIM along with 25,000 experimental data points. References to the citations that reported the experimental data are included. The SRIM-2003 Code in particular, allows the calculation of both longitudinal and lateral range and energy straggling of any energetic ion in different materials. The SRIM 2003, in particular, allows the calculation of both longitudinal and lateral range and energy straggling of any energetic ion in different materials.

1-7.3 Thin Target Yield, Effective Cross-Sections

Calculation of the thin-target yields y(E,0) were carried out as a function of projectile energy E. at the End of an Instantaneous Bombardment (EOIB), the yield $y(E)_{EOIB}$ is given by the slope at the origin of the growing curve of the activity per unit current (A/I) of a radionuclides vs. irradiation time, for a target in which the energy loss is negligible in respect to the projectile energy [cf. 3].

In practice, y(E) is defined from Eq. (1-30), as the second derivative of A/I in respect to particle energy and irradiation time, calculated when the irradiation time τ tends to zero (i.e., EOIB) [cf. 3]:

$$y(E)=y(E)_{EOIB}=y(E,0)=\left(\frac{\partial\left(\partial\left[A/I\right]\right)}{\partial E\,\partial\tau}\right)\quad with\,\tau\rightarrow 0\quad (Bq/C\,MeV),\qquad (1-30)$$

where A=Nλ is the radionuclides activity (Bq), I is the beam current (Amp.).

Experimentally, the function $y(E)_{EOIB}$ is calculated by the Eq. (1-31), that holds for a radionuclide produced by direct nuclear reaction only, without any decay charging and for very low dead counting times (i.e., LT≅RT; DT →0):

$$y(E)_{EOIB}=\left(\frac{C_\gamma}{(\varepsilon_\gamma I_\gamma LT)Q\Delta E}\right)D(RT)G(\tau)\exp(WT)\qquad (Bq/C\,MeV)\qquad (1-31)$$

where N the number of radioactive atoms, Q is the integrated proton charge (C) (obtained either from Faraday cup read-out or beam monitor reactions), C_γ is the net photo-peak counts at energy E_γ above background continuum, I_γ is the γ-emission absolute intensity, ε_γ is the experimental efficiency at the γ-energy considered, $\lambda = \ln(2)/T_{1/2}$ the decay constant ($sec.^{-1}$), LT is the live counting time (sec.), DT is the dead counting time (s), RT is the real counting time (sec.) =LT + DT ; WT is the waiting time from the EOB (sec.), t is the Irradiation Time (sec.), ΔE is the beam energy loss in the target (MeV) and the non-dimensional quantities D(RT) is the decay factor to correct decay during counting time and G(τ) is the growing factor to correct decay during irradiation, are defined as:

$$D(RT)=\frac{\lambda RT}{1\text{-}\exp(-\lambda RT)} \qquad (1-32)$$

$$G(\tau)=\frac{\lambda \tau}{1\text{-}\exp(-\lambda \tau)} \qquad (1-33)$$

Eq. (1-31) can be deduced from the Eq. (1-34) that defines the thin-target yield at the energy E:

$$y(E)=\frac{\sigma^{*}(E)N_{A}\lambda}{MZe^{-}\left(\left(dE/_{dS}\right)\cdot(E)\right)} \qquad \left(Bq/_{C\cdot MeV}\right), \qquad (1-34)$$

where M is the target atomic mass (g/mol), N_A the Avogadro's constant $=6.022045(31)\times10^{23}$ (atom/mol), E = <E> the ''average'' proton beam energy in the "thin" target (MeV), σ*(E) the "effective", or "weighed" (as explained below) reaction cross-section (cm^2), dE/dS the mass stopping-power (MeV/g/cm2), $e^{-}= 1.6022\times10^{-19}$ C, Z the atomic number of the projectile, S the mass thin-target thickness (g/cm^2).

In practice, the approximation ΔE≈S (dE/dS) was considered too crude for an accurate energy loss evaluation to be used in Eq. (1-34), thus ΔE was calculated by the difference of ranges of incoming and outgoing particle in the target from fitted proton range tables from [cf. **27**].

The "effective" cross-section σ*(E) (cm^2) as a function of projectile energy is defined by Eqs. (1-34), and (1-35) [cf. **3**]. Even if the physical meaning of this parameter is poor, being only a mere summation of the several cross sections $\sigma_i(E)$ of the reaction channels concerned, weighted on target isotopic composition (i.e. w_i and $\sum w_i = 1$), as obtained by the definition of Eq. (1-35):

$$\sigma^{*}(E)=\sum_{i} w_i \sigma_i(E), \qquad (1-35)$$

1-7.4 Thick-Target Yield Calculation

The thick-target yield Y(E, ΔE) is defined as a two parameter function of incident particle energy E(MeV) on the target and energy loss in the target itself ΔE (MeV) by Eq. (1-36). It holds in the approximation of a monochromatic beam of energy E; not affected by either intrinsic energy spread or straggling:

$$Y(E,\Delta E)=\int_{E-\Delta E}^{E} y(x)dx, \qquad (1-36)$$

in which the integrand y(x) represents the thin-target excitation functions of Eqs. (1-31), and (1-34). In case of total particle energy absorption in the target (i.e., energy loss ΔE=E), the function Y(E, ΔE) reaches a value $Y(E,E-E_{th})$; for $\Delta E = E-E_{th}$; that represents mathematically the envelope of the Y(E, ΔE) family of curves. This envelope is a monotonically increasing curve, never reaching either a maximum or a saturation value, even if its slope becomes negligible for high particle energies, and energy losses.

Eq. (1-36) states that the production yield of a thick-target does not increase further, if the residual energy in the target is lower then the nuclear reaction energy threshold, E_{th}. In practice, the use of a target thickness larger than the "effective" value is unsuitable from technological point of view, due to the larger power density Pd(W/g) deposited by the beam in target material itself, instead of target cooling system [cf. **3**].

The expected yield Y of a radioisotope from a particular thickness of target and nuclear reaction can be calculated using Eq. (1-37):

$$Y=\frac{N_A F}{M} I_b\left(1-\exp(-\lambda t)\right)\int\left(\frac{dE}{dx}\right)^{-1}\sigma(E)dE \qquad (1-37)$$

In this equation F is the fraction of the target isotope of the element, M is the molecular weight, I_b is the beam current, t is the irradiation time, λ is the decay constant of the produced nuclide, dE/dx is the stopping power, and σ(E) cross section correspond to energy E.

As the incident beam propagates through a thick target, its energy decreases due to the stopping power of the material. The reaction cross-section values in this case varying with the energy degradation. Supposing those values steady within successive intervals of thickness 1 MeV corresponding to stopping power of the target, the calculation of the yield is performed by integration over the whole energy range for the multiplication of the resulting stopping powers by cross section values.

The calculated value represents the maximum yield which, can be expected from a given target. In practice the experimentally obtained yield in high-current production runs is always lower than the theoretically calculated value, due to radiation damage effects, inhomogeneity in incident beam, losses of target material caused by the intense ion beam. The accurate knowledge of excitation functions, and therefore, the theoretical thick target yield, helps in designing target systems capable of giving optimum yields [cf. **17**].

1-8 The Aim of the Present Study

The aim of the present work is to study the excitation functions of the proton induced reactions on natural Mo and Gd, the radioactive products obtained from the physical processes involved in charged particle irradiation of individual isotopes. The experience gained from this study has been helpful in describing the adopted method in our laboratories to produce ^{94g}Tc, ^{95g}Tc, ^{95m}Tc, ^{96mg}Tc, $^{152mg}Tb$, ^{154m}Tb, ^{154g}Tb, ^{155}Tb, ^{156}Tb, and ^{160}Tb radionuclides via proton-cyclotron irradiation on molybdenum, gadolinium targets of natural isotopic composition. The development of routine methods for radioactively labeled precursor production for diagnosis and therapy as well as the development of procedure for the measurement of charged particle radiation. The technical aspects related to the uses of produced radionuclides for thin layer activation technique, and readjusting of nuclear data used Mo as a monitor are included in the study. The need for this study arose from the increasing number of medical applications that use ^{99m}Tc and its analogues ^{94}Tc, ^{95}Tc and ^{96}Tc radionuclides. Several nuclear tools are used to produce these radionuclides. Neutron induced reactions in reactors as well as proton induced reactions in cyclotrons and accelerators are used in the production of these radionuclides from natural and enriched elements, respectively. Small sized cyclotron is a promising means in the production of high purity radioisotopes with high specific activity. These small sized cyclotrons are of low energy range from 5 up to 20 MeV. The reaction cross sections of the studied reactions were determined experimentally and validated using nuclear model calculations.

To verify published experimental cross section data sets measured earlier, a new set of experimental thin-target excitation functions and "effective" cross-sections for direct nuclear reactions are performed, with incident proton energies in the range from threshold up to 18 MeV. New experimental observations and data might help theoreticians in developing some models for a concise description of the data. Dealing with one of the factors affecting the accuracy of measured data that is due to flux determination. This new approach revealed a new effect that clarified some aspects of the beam broadening due to straggling, and scattering effects were observed as a non-uniform increase in the beam penetration region. The multi monitors were then utilized in a new method for measuring charged particle beam fluxes. Thick-target yield values were calculated and optimized, by numerical fitting and integration of the measured excitation functions. These values allow optimization of production yield of one radionuclide, minimizing at the same time the yield of the others.

To perform this study, the available cyclotron and the associated facilities at Inshas Cyclotron Facility matched with the above mentioned requirements. The fundamental data of nuclear reactions useful for their production should be studied. For low energy cyclotrons (E < 20 MeV) the proton induced reactions on selected targets were chosen to produce these isotopes. The study includes cross section measurements of the investigated reactions. The activation method and the stacked foil technique using high resolution HPGe gamma ray spectrometer were applied to determine the excitation function.

Reliable data sets were produced with the help of simultaneous measurements of the excitation functions of $^{nat}Ti(p,x)^{48}V$, $^{nat}Cu(p,x)^{63}Zn$, and $^{nat}Cu(p,x)^{65}Zn$ monitor reactions. From the measured cross-sections the excitation functions are drawn and the integral yields are estimated for the entire product radionuclides. To validate the obtained cross section data nuclear model calculations of the same reaction cross sections are performed. The computer codes ALICE-91 [cf. **29**] and EMPIRE-II [cf. **30**], have been used in this work. The resulting cross-sections by the codes calculation are compared with the measured values for comparison.

2-1 Literature Review

2-1.1 Technetium-99m

The element of Technitium has isotopes with half-lives ranging from six hours up to sixty day, the decay schemes of these isotopes of most importance. Table (2-1) gives the decay data of radioisotopes of technetium and their gamma ray energies. ^{94m}Tc, ^{95m}Tc, ^{95g}Tc, and $^{96m,g}Tc$ have a convenient half-lives and a reasonable γ-ray energies. Many Tc radioisotopes have been artificially synthesized and sometimes chemically separated from irradiated targets. Some Tc isotopes can be used as general-purpose radiotracers to study the physicochemical properties of the element. Among them, ^{99m}Tc is at present, without any doubt, the most extensively used radiotracer in Nuclear Medicine imaging. In fact the short-lived ^{99m}Tc, which is commercially available from the decay of its parent radionuclide ^{99}Mo ($T_{1/2}$ = 2.7477 d), is widely used to label a wide range of organic complexes and coordination compounds for diagnostic purposes since (1956), when the generator was developed at Brookhaven National Laboratory [cf. **31**-**33**].

In more recent years, Nuclear Medicine has shown a growing interest in the cyclotron-produced short-lived positron-emitter ^{94m}Tc, as a flow agent bridging Single Photon Emission Computerized Tomography (SPECT) and Position Emission Tomography (PET) imaging [cf. **34**-**36**]. This last radionuclide seems promising for human PET investigations [cf. **37**-**41**] in spite of low spatial resolution, due to its high energy positron emissions (End Point=3460 keV, <753> keV, Firestone et al, (1996) [cf. **4**]. One aim of the present investigation is to define whether the very important medical radionuclide ^{99m}Tc and/or the $^{99}Mo–^{99m}Tc$ generator, can be practically produced using an alternative technology such as accelerator activation of Mo with protons.

Radioisotopes of technetium are of considerable interest in medicine because of their suitable nuclear and chemical characteristics. Technetium isotopes could label Organic and inorganic compounds due to multidentate coordination bonds according to chelation [cf. **42**,**43**]. During the early years of the last decade the available γ-cameras were suitable only for the low-energy rather than for the higher energy γ-rays. Therefore more efforts have been devoted to the production of ^{99m}Tc, and its analogues with another root rather than reactor by using cyclotrons. Furthermore, the Auger electrons emitted in the decay of $^{96m,g}Tc$ are of some therapeutic interest. Experimental thin-target excitation functions shall lead to deduce the "effective" cross-sections. Thick-target yield values calculated from the obtained data will be helpful in:

(1) Optimizing the production yield of one radionuclide,
(2) Ease in radiochemical separation of NCA (no carrier added) radionuclides,
(3) Minimizing at the same time the yield of the others by numerical fitting and integration of the measured excitation functions,
(4) High radionuclidic purity.

Reliable data sets can be produced with the help of simultaneous measurement of the excitation functions of $^{nat}Ti(p,x)^{48}V$, $^{nat}Ni(p,x)^{57}Ni$, $^{nat}Cu(p,x)^{62}Zn$, $^{nat}Cu(p,x)^{63}Zn$, $^{nat}Cu(p,x)^{65}Zn$ monitor reactions in the whole investigated energy range. The new cross-section values indicate the necessity of normalizing the excitation functions of about 250 of proton induced nuclear reactions by a factor of 0.8 measured earlier by V.N. Levkowski using the $^{nat}Mo(p,x)^{96mg}Tc$ process as monitor reaction. $^{96m,g}Tc$ is a gamma emitter and its half-life is suitable for long-term physiological investigation using Single Photon Emission Computed Tomography (SPECT).

Excitation functions for the $^{100}Mo(p,2n)^{99m}Tc$, $^{100}Mo(p,pn)^{99}Mo$ and $^{98}Mo(p,\gamma)^{99m}Tc$ nuclear reactions which are responsible for the production of ^{99m}Tc from natural element. The optimum energy range for the $^{100}Mo(p,2n)^{99m}Tc$ reaction is 22 to 12 MeV with a peak at ~17 MeV and maximum cross section of ~200 mb as suggested by Scholten et al., see [cf. **44**]. Over this energy range the production yield of ^{99m}Tc amounts to 11.2 mCi (415 MBq)/μA h or 102.8 mCi (3804 MBq)/ μA at saturation. There is no serious radionuclidic impurity problem. Production of ^{99}Mo is not viable due to the low cross section (~1.3 mb) over the proton energy range of 30 to 50 MeV. The $^{98}Mo(p,\gamma)^{99m}Tc$ reaction was found to have a cross section of <0.2 mb over the proton energy range studied.

The earlier assignment of this reaction channel as a potential route for production of ^{99m}Tc with a medical cyclotron is explained in [cf. **44**]. The countries that do not have access to fission molybdenum-99 for the $^{99}Mo \rightarrow ^{99m}Tc$ generator, could consider use of protons energy higher than >17 MeV at the cyclotrons for regional production of ^{99m}Tc. The $^{99}Mo \rightarrow ^{99m}Tc$ generator and ^{99m}Tc radiopharmaceuticals account for over 80% of nuclear medicine procedures worldwide. It is estimated that in the United States, Europe and Japan over 20 million nuclear medicine scans are performed annually that involve application of ^{99m}Tc [cf. **42**,**43**]. The present study was undertaken because of the continuing concern about the future supply of fission molybdenum-99 which is produced by nuclear reactors.

The feasibility of direct production of Curie quantities of ^{99m}Tc with a compact medical cyclotron by the ^{100}Mo(p,2n)^{99m}Tc nuclear reaction and lesser quantities of ^{99}Mo for the ^{99}Mo$\rightarrow$^{99m}Tc generator by the ^{100}Mo(p,pn)^{99}Mo nuclear reaction was reported by Beaver and Hupf (1971) [cf. **46**], and confirmed by Almeida and Helus (1977) [cf. **47**]. The investigators used Mo foils of natural isotopic composition and extrapolated the radionuclidic yields to a target of 97.4% isotopic enrichment in ^{100}Mo. The need for measurement of the excitation functions, production yields and resultant radionuclidic purity, as controlled by the isotopic composition of the target, was identified by Lambrecht et al. (1988) at an IAEA-Consultants Meeting on Nuclear Data for Medical Radioisotope Production [cf. **48**]. Subsequently, Lagunas-Solar et al. (1991); (1993) reported data using protons on Mo foils [cf. **49**,**50**].

Another data set was reported by Levkowski (1991) [cf. **51**], as mentioned by Schopper (1993) [cf. **52**], but no details are available. Lagunas-Solar et al. reported puzzling results (1991) and interpretation (1993) of the cross section data measured on Mo of natural isotopic composition. They were extrapolated to predict the excitation functions that would be expected to yield Tc radionuclides resulting from the proton irradiation of isotopically enriched Mo isotopes. There were considerable discrepancies between their prediction of the optimum proton beam energy and expected thick target yield and the earlier reports of Beaver and Hupf (1971), and Almeida and Helus (1977) [cf. **46**, **47**]. All the three groups had used Mo of natural isotopic composition for their irradiations, and then extrapolated the results to ^{100}Mo of 97.4% isotopic enrichment. The significant discrepancies in the nuclear data from the various laboratories aroused both interest and concern whether accelerator production of ^{99m}Tc and/or ^{99}Mo was a viable option for routine production of these important medical radionuclides. This is an important consideration for technologically developing countries or countries contemplating replacement of aging nuclear reactors for the purpose of medical radionuclide production.

Table (2-1): Principal Direct Nuclear Reactions, Decay Data Leading to the Main Radionuclides we are Interested in, that Occur in Proton Irradiated ^{nat}Mo and their Calculated Threshold Energies, E_{th} [cf. **45**]. All the Radionuclides Cited are Identified in the γ-Spectra of this Experiment, either at EOB or some Time Later [cf. **4**].

Nuclide	Half-Life	Decay Mode (%)	Eγ (MeV)	Iγ (%)	Contributing Reactions	Threshold (MeV)
^{99m}Tc	6.01h	IT(100)	140.5	89.06	^{100}Mo(p,2n) ^{98}Mo(p,γ) ^{100}Mo(p,pn)	7.79 0.0 6.13
^{99}Mo	65.94h	β^-(100)	181.0	6.07	^{100}Mo(p,pn)	6.13
^{96m}Tc	51.5m	EC(98)	778.2 1200.2	1.9 1.08	^{96}Mo(p,n) ^{97}Mo(p,2n) ^{98}Mo(p,2n)	3.79 10.69 19.42
^{96g}Tc	4.28d	EC(100)	778.2 812.5 849.9 1126.8 1200.2	99.76 82.0 98.0 15.0 0.37	^{96}Mo(p,n) ^{97}Mo(p,2n) ^{98}Mo(p,3n) ^{96m}Tc decay	3.79 10.69 19.42
^{96}Nb	23.4h	β^- (100)	568.8 778.2 849.9 1200.2	58.0 96.45 20.45 19.97	^{96}Mo(p,2p) ^{97}Mo(p,2pn) ^{98}Mo(p,pd) ^{98}Mo(p,^{3}He) ^{100}Mo(p,αn)	9.32 18.05 15.80 10.25 3.82
^{95m}Tc	61.0d	EC(96.12)	835.1	26.6	^{95}Mo(p,n) ^{96}Mo(p,2n) ^{97}Mo(p,3n)	2.50 11.75 18.64
^{95g}Tc	20.0h	EC(100)	765.8	93.8	^{95}Mo(p,n) ^{96}Mo(p,2n) ^{97}Mo(p,3n) ^{95m}Tc decay	2.50 11.75 18.64
^{94m}Tc	52.0m	EC(100)	1868.7	5.7	^{94}Mo(p,n) ^{95}Mo(p,2n) ^{96}Mo(p,3n)	5.09 12.54 21.56
^{94g}Tc	4.9h	EC(100)	702.7	99.7	^{94}Mo(p,n) ^{95}Mo(p,2n) ^{96}Mo(p,3n) ^{94m}Tc decay	5.09 12.54 21.56
^{92m}Nb	10.15d	EC(100)	934.4	99.07	^{94}Mo(p,^{3}He) ^{95}Mo(p,α) ^{96}Mo(p,αn)	9.71 0.0 5.61
^{48}V	15.97d	EC(89), β^+(11)	944.1 983.5	7.76 99.98	^{nat}Ti(p,x)	4.89
^{57}Ni	35.6h	EC+β^+(100)	127.2 1377.6	16.7 81.7	^{nat}Ni(p,x)	10.17
^{62}Zn	9.168h	EC+β^+(100)	548.3 596.6	15.3 26.0	^{nat}Cu(p,x)	13.47
^{63}Zn	38.47m	EC+β^+(100)	669.6 962.1	8.0 6.5	^{nat}Cu(p,x)	4.21
^{65}Zn	244.26d	EC+β^+(100)	1115.5	50.6	^{nat}Cu(p,x)	2.17

While $^{99}Mo \rightarrow ^{99m}Tc$ generator is usually produced in nuclear reactor by both thermal neutron irradiation of natural $^{98}Mo(n,\gamma)^{99}Mo \rightarrow ^{99m}Tc$ and thermal fission on highly enriched ^{235}U via $^{235}U(n,f)^{99}Mo$-processes Boyd, (1983) [cf. **33**]. ^{94m}Tc is more commonly produced by accelerators via irradiation by protons, deuterons, helium-3 and alpha beams of either ^{93}Nb or ^{94}Mo enriched targets [cf. **38**,**39**,**40**,**53**-**55**]. Up to now, less effort in this area was devoted to production of other very short-lived Tc radionuclides, such as ^{92}Tc and ^{93g}Tc, even if their application in the Nuclear Medicine imaging field by PET, seems promising, as suggested by [cf. **38**,**56**].

2-1.2 Technetium-99m Analogues

The ^{99m}Tc analogues that could be produced from natural Mo target via irradiating molybdenum targets of natural isotopic composition by protons are ^{94g}Tc, ^{95g}Tc, ^{95m}Tc and ^{96g}Tc radionuclides [cf. **3**].

In the literature, several experimental data and theoretical calculations are available, which describe either the nuclear cross-sections or the thin-target yields of Tc isotopes from targets of natural isotopic composition, in a sufficiently wide energy range [cf. **56**-**58**]. The studies were mostly carried out with proton [cf. **58**-**63**], deuteron [cf. **54**], alpha [cf. **41**,**53**,**65**], ^{3}He [cf. **38**] beams. In particular, alpha and helium-3 irradiation on ^{93}Nb is justified from basic nuclear physics point of view by the mono-isotopic composition of natural niobium.

Some papers were published on photo-nuclear reactions on ^{99g}Tc [cf. **64**,**65**] deal with irradiation of ^{99g}Tc to produce Ru radionuclides, reporting data concerning, also the production of $^{95,96,99m}Tc$ via side reactions [cf. **66**]. For production purposes proton beams are normally preferred [cf. **3**], due to the lower stopping-power and larger range than deuterons and helions, as calculated by the SRIM-2003 code [cf. **27**]. The use of variable energy proton-cyclotron beams presents remarkable advantages to induce (p,xn) reactions on either natural or isotopically enriched targets with the aim of determining excitation functions and cross-sections. For this reason, we engaged a set of experimental determination of the thin-target excitation functions and the “effective” cross-sections for the nuclear reactions $^{nat}Mo(p,xn)^{A}Tc$ (with A = 94, 95, 96 and 99) [cf. **4**].

Production via irradiation of Mo is very advantageous due to the low cost of natural metallic molybdenum, its good thermal and electrical conductivity and its very high melting point (2623 °C) [cf. **4**]. Some literature works comparable with the present we do exist, but were carried out with different aims. They report only some thick-target yield values at specific energies for wear studies Ouellet et al., (1993) [cf. **67**], or consist of systematic studies, but in a lower energy range Roesch and Qaim, (1993) [cf. **39**]. Besides the direct (p,xn) reactions, other side reactions that produce some Mo, Nb and Zr radionuclides occur in the ^{nat}Mo target [cf. **44**,**49**,**50**,**51**].

Recommended cross-sections for monitor reactions induced with proton to produce $^{96m,g}Tc$, and ^{96}Nb by the reactions $^{nat}Mo(p,x)^{96m,g}Tc$; $^{nat}Mo(p,x)^{96}Nb$ processes also are of a great benefit in TLA (Thin Layer Activation Technique). Cumulative cross-sections, thick target yields and activation functions would be deduced and compared with available literature data. Molybdenum targets of natural abundance used for proton induced reactions to obtain the cumulating, and isomeric cross-sections ratios, and thick target yields and the excitation functions aid in assigning activation curves can be used in the field of TLA-technique and beam monitoring for proton beam.

2-1.3 Terbium

Our research interests for Terbium is related to the field of Medicinal Inorganic Chemistry, in particular the chemistry of complexes of the lanthanides, such as terbium, gadolinium, and europium as luminescent sensors. Tb complexes could be used as contrast agents in MRI (magnetic resonance imaging), which visualizes water molecules in the human body We are also interested in the use of radioisotopes of certain lanthanides and group 13 metals in diagnostic and therapeutic nuclear medicine. Several terbium complexes in macroscopic amounts are used in medical research as contrast agents for Magnetic Resonance Imaging (MRI) [cf. **68**, **69**].

In order to study the bio-distribution of those magneto-pharmaceuticals *in vivo* experiments were done on animals using ^{152m}Tb ($T_{1/2}$ = 17.5 h) as a radionuclidic marker [cf. **70**,**71**]. Due to its short half-life, this radioisotope, however, terbium could be used in humans rather than gadolinium ^{153}Gd ($T_{1/2}$ = 241.6 d) of longer half-life which, cannot be used. To study the detailed behavior of MRI contrast agents, it would be meaningful to combine the techniques of MRI and Single Photon Emission Computed Tomography (SPECT) by using an appropriately labeled magnetopharmaceutical [cf. **72**,**73**]. The data obtained on the uptake kinetics would allow a better estimate of the diagnostic potential of new Tb-compounds in MRI. Of all the radioisotopes of terbium, ^{152m}Tb seem to be the most suitable for SPECT imaging: it has a half-life of 17.5 h and a strong γ-ray of 344 keV (65%). ^{160}Tb is of considerable importance due to its decay mode by electron emission of 209 keV (100%) as well as a cascade of auger electrons accompanying its γ-ray emission [cf. 74,75]. Table (2-2) gives the decay data of radioisotopes of technetium and their gamma ray energies. ^{154m}Tb, ^{154g}Tb, ^{155}Tb, ^{156}Tb, and ^{160}Tb have a convenient half-lives and a reasonable γ-ray energies.

The water present in various tissues behaves differently, allowing a detailed, non-invasive image of the body to be generated. Signal intensity is largely dependent on the time excited water molecules take to relax back to their normal states. When in close contact with certain metal ions (e.g. gadolinium) this can be rapid. Terbium complexes (contrast agents) are thus injected into the body to enhance image contrast [cf. 72,73].

The slower the Terbium complex tumbles in solution, the faster it relaxes giving a more intense signal. One method of slowing the tumbling has been the formation of host-guest non-covalent interactions between the Terbium chelate and slowly tumbling macromolecules. The most successful applications to date have been the design of chelates bearing hydrophobic units that can bind to the protein Human Serum Albumin (HSA). By careful design of novel hydrophobic moieties and rigid spacers, aims to produce complexes that remain in the blood stream and image the cardiac region. Such compounds are being sought for use in Magnetic Resonance Angiography [cf. 76]. All the Radionuclides Cited are Identified in the γ-Spectra of this Experiment, either at EOB or some Time Later from the decay data that are taken from the Table of Isotopes [cf. 4].

A review of the past decades research into lanthanides complexes as MRI contrast agents reveals a paucity of other examples than terbium, with most research concentrating on subtle variations on these themes. In light of this, we believe there is scope to develop new types of ligand architectures for the lanthanides [cf. 77]. By reviewing the computer index for experimental nuclear data library for proton induced reaction that could be carried out on gadolinium target we did not find any available for ^{152,154,155,156,157,158,160}Gd. So this is a fact reaction explain itself the importance to carry out such a measurements to find out this route of reactions that can lead to produce some radionuclides of most importance in several technological purposes. For example to produce ^{152,154,155,156,160}Tb From the isotopes ^{nat}Gd only one work done by C. Birattari et al., 1973 [cf. 78], for (p,n) reaction cover the energy range from 3.4-43.6 MeV. We recommand the use of terbium radioisotopes in nuclear medicine for therapy and diagnosis, due to its decay scheme properties.

The significant discrepancies in the nuclear data from the various laboratories aroused both interest and concern whether accelerator production of ^{99m}Tc and/or ^{99}Mo was a viable option for routine production of these important medical radionuclides. This is an important consideration for technologically developing countries or countries contemplating replacement of aged nuclear reactors for the purpose of medical radionuclide production. Excitation functions were measured by the conventional stacked-foil technique Piel et al., (1992); Scholten et al., (1995); Takacs et al., (2002) [cf. **79,80,81**].

Table (2-2): Principal Nuclear Reactions, Decay Data Leading to the Main Radionuclides We Are Interested in, that Occur in Proton Irradiated ^{nat}Gd and their Calculated Threshold Energies, E_{th} [cf. **4,45**].

Nuclide	Half-Life	Decay Mode (%)	Eγ (MeV)	Iγ (%)	Contributing Reaction	Threshold (MeV)
^{152}Tb	17.5h	EC(100)	344.4 596.3	65.0 9.4	^{152}Gd(p,n)	4.80
^{154m}Tb	9.4h	EC(78.2)	247.9 540.1 649.5 996.3 1004.7	22.0 20.0 11.0 9.0 11.0	^{154}Gd(p,n) ^{155}Gd(p,2n) ^{156}Gd(p,3n)	4.37 10.85 19.44
^{154g}Tb	21.5m	EC(100)	722.1 996.3 1274.4	8.0 5.0 11.0	^{154}Gd(p,n) ^{155}Gd(p,2n) ^{156}Gd(p,3n) ^{154m}Tb decay	4.37 10.85 19.44
^{155g}Tb	5.32d	EC(100)	86.55 105.32 180.1 262.3	31.81 24.91 7.4 5.2	^{154}Gd(p,γ) ^{155}Gd(p,n) ^{156}Gd(p,2n) ^{157}Gd(p,3n)	0.00 1.61 10.21 16.61
^{156}Tb	5.35h	EC(100)	199.2 262.54 537.99	41.0 5.8 67.0	^{155}Gd(p,γ) ^{156}Gd(p,n) ^{157}Gd(p,2n) ^{158}Gd(p,3n)	0.00 3.25 9.65 17.64
^{160}Tb	72.3d	$β^-$ (100)	879.4 298.58 966.2 1177.96	30.0 25.51 25.21 15.1	^{160}Gd(p,n)	0.89

2-2 Summary of Earlier Investigations

Table (2-3) gives a summary of the earlier investigations of the proton-induced data from the available literature. From the literature survey, it is noticed that some excitation functions of proton induced reactions producing Tc and Tb were not well investigated, specially in the low energy range of 5 to 18 MeV.

The major objective of this work was to evaluate the potential of utilizing proton accelerators as a supply source of ^{99m}Tc and its analogues family as well as Tb new radionuclides for use in diagnostic and therapy nuclear medicine.

Table (2-3): Gives a Summary of the Earlier Investigations of the Proton Induced Data from the Literature.

Author [cf.]	Target	Irradiation	Monitor Reaction	Measurement of Activity	Decay Data (D) Stopping Power S)	Reaction Measured	Quantity No. of Data Points	Energy Range
Hogan [cf. **59**]	Mixture of Mo-Oxides, Cu-Oxides, Enrichment 96.8%	Cyclotron Stacked foil method	Cu-monitor	Ge(Li) no Chemical Separation	Not given	$^{96}Mo(p,x)^{96m}Tc$ $^{96}Mo(p,x)^{96g}Tc$	Sig. No.10 Sig. No.10	10.0-80.0 10.0-80.0
Almeida and Helus [cf. **47**]	^{nat}Mo, foils 150 μm thick	Cyclotron Stacked foil method	Faraday Cup, Cu-monitor	Ge(Li) Chemical Separation	Not given	$^{nat}Mo(p,x)^{99m}Tc$ $^{nat}Mo(p,x)^{96}Tc$ $^{nat}Mo(p,x)^{95}Tc$ $^{nat}Mo(p,x)^{94}Tc$ $^{nat}Mo(p,x)^{93}Tc$ $^{nat}Mo(p,x)^{99}Mo$	Yield No.7 Yield No.7 Yield No.7 Yield No.7 Yield No.7 Yield No.7	8.0-24.5 5.0-24.5 3.0-24.5 5.5-24.5 10.0-24.5 9.0-24.5
Skakun et al [cf. **81**]	^{96}Mo, 1-5 mg/cm^2	Linac, $E_{p=}9$ MeV	Faraday Cup,	Ge(Li) no Chemical Separation	Lederer and Shirley 1978 (D)	$^{96}Mo(p,x)^{96m}Tc$ $^{96}Mo(p,x)^{96g}Tc$	Sig. No.10 Sig. No.10	4.5-9.0 4.5-9.0
Lagunas-Solar [cf. **49**]	^{nat}Mo, foils 12.7 μm thick	Cyclotron Stacked foil method	Faraday Cup, Cu-monitor	Ge(Li) no Chemical Separation	Lederer and Shirley 1978 (D) Williamson et al., 1966 (S)	$^{nat}Mo(p,x)^{99m}Tc$ $^{nat}Mo(p,x)^{96}Tc$ $^{nat}Mo(p,x)^{95}Tc$ $^{nat}Mo(p,x)^{94}Tc$ $^{nat}Mo(p,x)^{99}Mo$	Sig. No.75 Sig. No.75 Sig. No.75 Sig. No.75 Sig. No.62	9.54-24.6 9.54-24.6 9.54-24.6 9.54-24.6 13.3-24.6
Levkowski [cf. **82**]	Oxide of Enriched ^{97}Mo, ^{98}Mo, ^{100}Mo,	Cyclotron Rotating target	Mo-monitor $^{nat}Mo(p,x)^{96}Tc$	Ge(Li) no Chemical Separation	Not given	$^{96}Mo(p,n)^{96mg}Tc$ $^{97}Mo(p,2n)^{96mg}Tc$ $^{98}Mo(p,3n)^{96mg}Tc$ $^{97}Mo(p,2p)^{96}Nb$ $^{98}Mo(p,2pn)^{96}Nb$ $^{100}Mo(p,\alpha n)^{96}Nb$	Sig. No.13 Sig. No.21 Sig. No.11 Sig. No.7 Sig. No.5 Sig. No.16	7.7-18.3 11.0-29.5 20.3-29.5 24.0-29.5 25.7-29.5 15.8-29.5
Zuravlev et al., [cf. **83**]	Enriched ^{96}Mo, 96.7 %	Cyclotron Stacked foil method	Faraday Cup,	Ge(Li) no Chemical Separation	Not given	$^{94}Mo(p,n)^{96g}Tc$ $^{95}Mo(p,n)^{95m}Tc$ $^{96}Mo(p,n)^{95g}Tc$	Sig. No.2 Sig. No.2 Sig. No.2	5.43-5.95 5.15-6.02 5.5-5.95

Table (2-3): Gives a Summary of the Earlier Investigations of the Proton Induced Data from the Literature, (Followed).

Author [cf.]	Target	Irradiation	Monitor Reaction	Measurement of Activity	Decay Data (D) Stopping Power S)	Reaction Measured	Quantity No. of Data Points	Energy Range
Lagunas-Solar [cf. **84**]	^{nat}Mo, foils 12.7 μm thick	Cyclotron Stacked foil method	Faraday Cup,	Ge(Li) no Chemical Separation	Lederer and Shirley 1978 (D) Williamson et al., 1966 (S)	$^{nat}Mo(p,x)^{99m}Tc$ $^{nat}Mo(p,x)^{96}Tc$ $^{nat}Mo(p,x)^{95}Tc$ $^{nat}Mo(p,x)^{94}Tc$ $^{nat}Mo(p,x)^{93}Tc$ $^{nat}Mo(p,x)^{99}Mo$	Sig. No.50 Sig. No.50 Sig. No.51 Sig. No.50 Sig. No.49 Sig. No.49	10.6-67.5 7.6-67.5 5.1-67.5 5.1-67.5 12.0-67.5 19.0-67.5
Wenrong et al., [cf. **85**]	^{nat}Mo, foils, 6-20 mg/cm^2 thick	Tandem VdG	Faraday Cup,	HPGe no Chemical Separation	Williamson et al., 1966 (S)	$^{nat}Mo(p,n)^{96g}Tc$ $^{nat}Mo(p,n)^{95m}Tc$ $^{nat}Mo(p,n)^{95g}Tc$	Sig. No.17 Sig. No.17 Sig. No.17	6.81-21.85 6.81-21.85 6.81-21.85
Scholten et al., [cf. **44**]	Enriched ^{100}Mo-Foil, ^{98}Mo-Oxide	Cyclotron Stacked foil method	Faraday Cup, Cu-monitor	HPGe no Chemical Separation	Firestone, 1996 (D) Brown and Firestone 1986 (D) Williamson et al., 1966 (S)	$^{100}Mo(p,2n)^{99m}Tc$ $^{98}Mo(p,\gamma)^{99m}Tc$ $^{100t}Mo(p,2p)^{99}Mo$	Sig. No.27 Sig. No.22 Sig. No.26	16.0-3.0 38.2-22.1 45.5-29.7 45.5-13.7 65.0-42.0
Bonardi et al., [cf. **3**]	^{nat}Mo, foils	Cyclotron single foil irradiation	Faraday Cup, Cu-monitor	HPGe no Chemical Separation	Brown and Firestone 1986 (D) Williamson et al., 1966 (S)	$^{nat}Mo(p,x)^{96mg}Tc$ $^{nat}Mo(p,x)^{95g}Tc$ $^{nat}Mo(p,x)^{95m}Tc$ $^{nat}Mo(p,x)^{94g}Tc$	Yield No.17 Yield No.17 Yield No.17 Yield No.17	4.9-43.7 4.9-43.7 4.9-43.7 4.9-43.7
Takacs et al., [cf. **86**]	^{nat}Mo, foils	Cyclotron Stacked foil method	Faraday Cup, Cu-monitor	HPGe no Chemical Separation	Brown and Firestone 1986 (D) Anderson Ziegler 1983 (S)	$^{nat}Mo(p,x)^{96mg}Tc$ $^{nat}Mo(p,x)^{96}Nb$	Sig. No.24 Sig. No.18	5.7-37.91 19.0-37.9
Birattari et al., [cf. **78**]	^{nat}Gd, foils 19 mg/cm^2	Cyclotron Stacked foil method	Faraday Cup,	Ge(Li) no Chemical Separation	Not given	$^{nat}Gd(p,x)^{160}Tb$	Sig. No.24	4.3-43.6

3-1 Introduction

This chapter contains the experimental procedures used in the preparation of stacks, irradiation conditions, gamma ray counting of the resulting radionuclides, beam energy, flux determination, the correction error considered through this study, and cross section measurements by using the activation technique. Before cross section determination, we have paid attention to the uncertainties in beam energy, detector absolute efficiency, and beam flux determination. These three quantities are of considerable importance, affecting the resultant cross section values. The exact incident energy and the effective detector efficiency were quite dealt by a settled technique. We introduced an approach to determine the incident flux more accurately. The Inshas Cyclotron facility that used for irradiation and measurement is demonstrated by schematic diagrams and detailed descriptions. The estimation of uncertainty and sources of experimental errors are given at the end of this chapter.

3-2 The Cyclotron

Ernest Lawrence first demonstrated the use of charged particle accelerators for radionuclide production in (1934) [cf. **9**]. The development of the fission nuclear reactor during World War-II, and the availability of the comparatively cheap and long lived fission products and neutron induced isotopes, resulted in a retardation of cyclotrons used in radionuclide production.

The cyclotrons are finding increased applications in the production of radionuclides, especially for medical purposes. Today a large number of cyclotrons are used worldwide for medical radioisotope production [cf. **15**,**87**]. The common terminology for these cyclotrons is "medical cyclotron" or "compact cyclotron". Many of them have been installed in hospital environments and are employed extensively for preparation of short-lived radionuclides with very high-specific activities for direct use on site. For reasons of cost and efficiency, cyclotrons are the most commonly used accelerators for some radionuclide production. More specifically, cyclotrons are technically less complicated than other accelerators and the obtained particle beam currents obtainable are high.

Level I cyclotrons, having rather low beam energy ≤10 MeV, often require the use of isotopically enriched target isotopes in order to produce sufficient amounts of the desired radionuclides. When using level II cyclotrons the targets turned to enriched target materials, mostly in order to avoid interferring reactions, resulting in a contamination of the radionuclidic product. The work presented in this thesis has been carried out with the 103 cm AVF (level II) cyclotron at Inshas, Nuclear Research Center, Atomic Energy Authority, Cairo, Egypt. Figure (3-1) displays a picture for the 103 AVF (MGC-20) cyclotron. A variable energy, multi-particle isochronous cyclic accelerator is installed. This compact machine is particularly suited for regional production of commercially distributing radioisotopes for clinical use. In Figure (3-2) a scheme of the layout of the laboratory is shown [cf. **88**]. The characteristics of the machine are summarized in Table (3-1). The scope of any radionuclide production program is the determining factor in cyclotron choice. Table (3-2) makes a relation between the program and the particles, particle energy, available current of the machine would be used in radionuclides production [cf. **89**]. While there can be an extension of scope of nuclide production in four particles cyclotron Level-II machine.

Table (3-1): Particles Species, Their Energies and Intensities for the External Beam of the Inshas Cyclotron MGC-20 [cf. **88**].

Particle	Energy Range [MeV]	Maximum Particle Beam Intensity [μA]
Proton	5-18	50
Deuteron	3-10	50
^{3}He	8-24	25
Alpha	6-20	25

Level-II machines can support a research program in addition to the clinical program. But due to the small quantities of acceptable pure radionuclides produced using ^{3}He and ^{4}He (in the 6-24 MeV) the machine may be of single use in research program more than in a clinical program. To produce ^{13}N, ^{15}O and ^{18}F (all p-induced) enriched stable isotopes must be used, beside the cost factors involved, efficient targetry and recovery systems have to be developed. Level-II machine has the flexibility for switching source gases, H_2, D_2 at the control console and automated control of the requisite RF. Thus particle switching and beam readjustment time should not exceed 5-10 minutes. Once the beam has passed through the vacuum isolation foil of the machine, the energy and the current bringing heat into the foil itself. Generally up to 2000 Watts presents no problem and the design of most system tolerating higher. The vacuum system of the machine must be protected with rapidly closing automated valves might the isolation in case of foil rupture.

Figure (3-1): Picture for the 103 AVF (MGC-20) cyclotron (level-II) at the Inshas Cyclotron Facility.

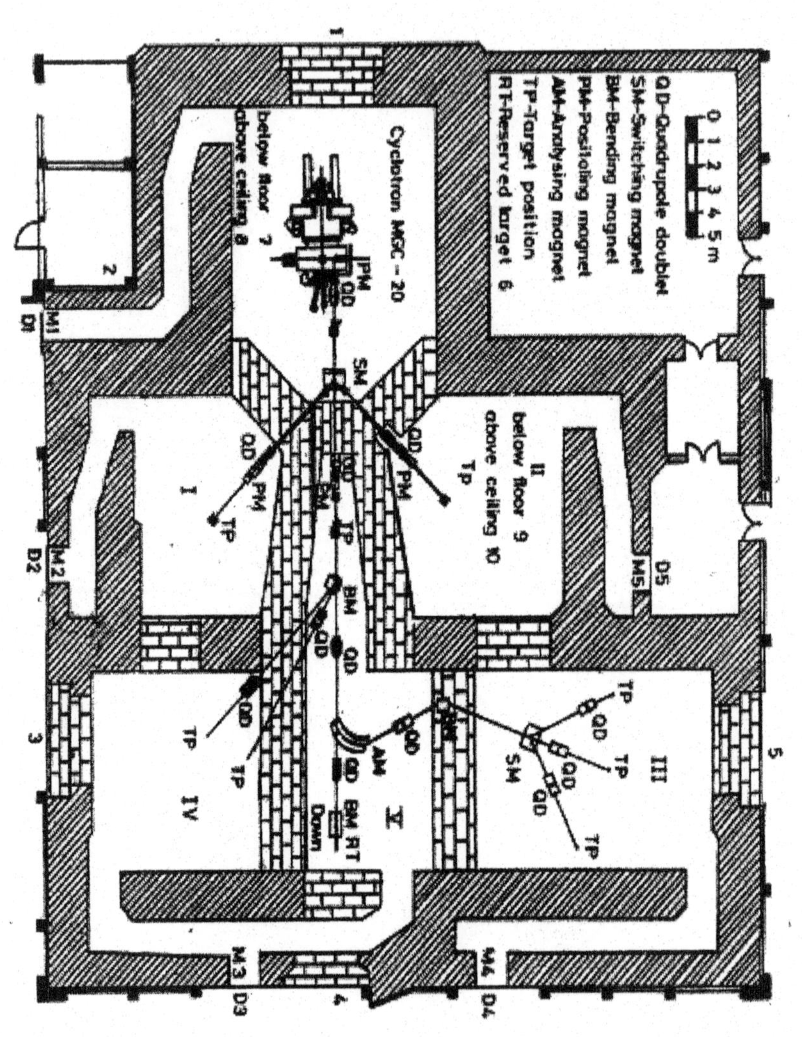

Figure (3-2): Layout of Cyclotron and Beam Transport System at the Cyclotron Laboratory of Inshas MGC-20 (AVF) [cf. **88**].

Table (3-2): Gives a Relation Between the Machine and the Particles, Particle Energy, Available Current.

Level II Cyclotrons							
Manufacturer	**Scand MC16F**	**JSWBC168**	**BC1710**	**CG12 Sum 325**	**CS-18**	**CP-18^**	**MGC-20**
Proton (MeV)	**17**	**1[illegible]**	**17**	**16**	**17**	**17**	**18**
Deuteron (MeV)	**8.5**	**[illegible]**	**10**	**8**	**9**	**9**	**10**
Other avail. Particle	**yes**	**n[illegible]**	**no**	**no**	**yes**	**no**	**yes**
Proton Current (μA)	**50**	**5[illegible]**	**50**	**50**	**60**	**50**	**50**
Deuteron Current (μA)	**50**	**5[illegible]**	**50**	**50**	**100**	**-**	**50**
Weight (Kt)	**17Kt**	**> [illegible]**	**> 33**	**> 14**	**23**	**23**	**24**
Ion Source	**axial**	**axial-penning**	**axial-penning**	**radial-Living-stone-Jones**	**radial-penning**	**radial-penning**	**axial-penning**
Dee System	**2 x 90°**	**2 x [illegible]5**	**2 x 45**	**1x 180**	**2 x 90**	**-**	**2 x 140**
Mag. Field Stg. Tesla	**-**	**1.5[illegible]**	**1.4p, 1.5d**	**1.78**	**1.75**	**-**	**1.4**
Accel. Frequency, MHz (p)	**-**	**47**	**43.5**	**26**	**27**	**-**	**-**
Accel. Frequency, MHz (d)	**-**	**47**	**47**	**40.5**	**14**	**-**	**-**
Pumping	**Diffusion**	**Turbomolec.**	**Diffusion**	**Diffusion**	**Oil-Diffusion**	**Oil-Diffusion**	**Diffusion**
Installed Power	**110 KVA**	**183KVA**	**-**	**150KVA**	**120KVA**	**-**	**260KVA**
Self Shielding	**Avail.**	**Partial**	**partial**	**none**	**Avail.,**	**none**	**none**

All machines listed are isochronous AVF, MC16F–Scanditronix, Upsala, Sweden; JSWBC-168 and BC-1710 Japan Steel works, Muroran, Japan; CG-12 and Sum-325, CGR-MeV-SUMITOMO, Buc, France, Tokyo, Japan; CS, CP The Cyclotron Corporation, Berkeley, CA, USA. (^ Variable energy negative ion machine) [cf. 89].

The beam shape emerging from a cyclotron usually is roughly rectangular, some shaping may be introduced inside the cyclotron by the use of magnetic channels, and exterior manipulation of the beam to produce an optimum shape. To maximize the use of rectangular beam targets constructed with the same shape, advantages of the cylindrically or conically shaped targets are so mandate and concordant. Using of focusing magnet systems can shape beams, the complexity of the system determine the quality of the shape. Usually an elliptical shape with relatively short semi-major axis (cross section from 0.5 to 4.0 cm sometime to 10 cm) is desirable. Since the optimization of the yield from particular target depends on the chemistry of the target which is sensitive to both Dose/Unit volume and Dose rate/Unit volume during irradiation. That providing focusing, expansion or contraction of the beam. The beam should also have a uniform density in the cross-section, lake of uniformity, especially the presence of hot spots (i.e. very small areas of beam where the particle flux can be 10-10^4 higher than the average beam), can cause repeated foil ruptures. Where heat transfer from these microscopic areas is not sufficiently rapid. This becomes especially important in high-pressure target systems. The solution is in the use of rapid oscillation (500-1000 Hz) of a small beam in a spiral or raster pattern to improve window performance.

3-3 Stacked Foils Technique

Studying the experimental behavior of the excitation functions for Tc, and Tb radionuclides, thin Mo, and Gd foil targets of uniform thickness were irradiated with protons of energies 15, and 18 MeV. The proton energy degradate through each foil was about 500 keV. High purity Gd (99.9%) metallic foils (Goodfellow Metals, UK), and high purity Mo (99.9%) metallic foils (Sigma-Aldrich Chemie, Germany) have been used. The thickness uniformity of the foils were determined by weighting of a large piece of foil having a well defined surface area. They were 10.0μm±2.0% for Gd, equivalent to 7.901 mg/cm^2±2.0%, and 25.0μm±1.4% for Mo, equivalent to 25.7 mg/cm^2±1.4%. Table (3-3) contains the isotopic composition of natural Molybdenum, and Gadolinium.

In order to obtain the excitation function of any reaction, a number of cross section values over the energy range under investigation need to be measured. It is essential to use thin samples so that minimum energy degradation in each sample is achieved. On the other hand, the sample should be thick enough to produce measurable activity to be able to obtain good counting statistics.

Table (3-3): Isotopic Composition of Natural Molybdenum, Gadolinium Used in this Study [cf. **4**].

Isotopic Mass/Isotopic Abundance (%)						
^{92}Mo	^{94}Mo	^{95}Mo	^{96}Mo	^{97}Mo	^{98}Mo	^{100}Mo
14.84	9.25	15.92	16.68	9.55	24.13	9.63
^{152}Gd	^{154}Gd	^{155}Gd	^{156}Gd	^{157}Gd	^{158}Gd	^{160}Gd
0.20	2.18	14.8	20.47	15.65	24.84	21.86

Excitation functions of the concerned reactions were measured by the conventional stacked-foil technique [cf. **79,80,81**]. Figure (3-3) shows a schematic diagram for stacked foils arrangement. This technique allowing the examination of many target elements at different energy values in one run with a minimum requirement of beam-line time. The simultaneous irradiation of the samples (the stack) is not only economical but also could assure a good relative accuracy.

The stacked target technique could be applied for targets that consist of thin foils. Among the stack target foils one or more monitor foils are inserted for particle beam monitoring. In special cases energy degrade foils are placed in the stack, their purposes are to be used as a catcher for recoilled nuclei as well as to enhance the energy degradation in the stack. Otherwise, to cover the whole energy region a large number of target foils must be necessary. Every sample sandwiched between to monitor reaction foils of Copper, and Titanium in one case. In the other case the two monitor foils putted as a couple separate every two successive samples. After the stacked foils have been irradiated, the gamma rays spectrometer was used for counting the samples. The targets were irradiated with the proton beam at the AVF (Azimutally Variable Field) Cyclotron of the Inshas Cyclotron Facility, using the "stacked foils" activation technique for one hour.

The targets were all set in the same geometric configuration, with a beam current of about 100-200 nA and an integrated charge varying from 375 to 750 mC, measured with a long shaped Faraday cup. The beam intensity was corrected by the monitor reactions at every sample target foil, obtaining the exact flux within the 2%. The cyclotron maximum extracted beam energy was 18.0±0.2 MeV; with an intrinsic energy spread of ±0.05 MeV. Two irradiation facilities have been used. In the first the target holder stand at the end of beam line inside the vacuum, in the second a 25 μm thick tantalum window was used to separate the beam line from irradiation target holder that was irradiated in atmosphere. The beam energy was degraded from the nominated value to reaction thresholds, at about 1.0 MeV intervals through the stack. Calibrations were used to set the energy [cf. 27]. The beam energy was calibrated with a ±0.2 MeV accuracy, by the activity ratios of two reactions generated simultaneously in Cu foils.

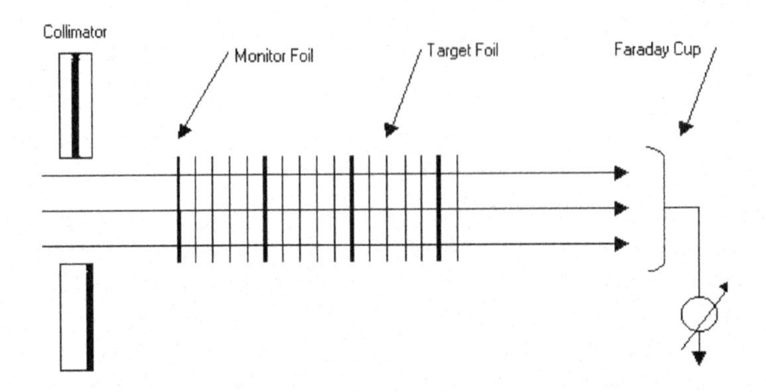

Figure (3-3): Stacked Foils Technique Arrangement.

Systematic errors related to both stopping-power calculations and absolute γ-emission intensities were not considered. The high-resolution gamma-spectrometry measurements were carried out in geometrical conditions (not less than 20–25 cm distance from the detector) suitable for minimizing any counting loss due both to counting dead time and random pile up [cf. 90,91,92]. In the case of ^{94g}Tc, the dead time (DT) was of the order of 10%, which is acceptable for an accurate correction, by commercial electronics (EG&G, Ortec 570 spectroscopy amplifier), without use of constant frequency pulse generator correction method. In case of ^{96}Tc the cumulative yield, and "effective" cross-sections σ* of both metastable and ground levels are reported.

3-4 Calibration of Cyclotron Beam Energy

The primary beam incident energies were nominated from the cyclotron parameters, and derived from our calibration measurements based on the activation technique according to Sonck et al., (1997); Tarkanyi et al., (2001) [cf. 93,94]. Since we applied a stacked-foil technique we had to calculate the proton energies in all the different target foils. Although the energy degradation was small for the highest initial proton energies we performed such calculations for all irradiations. Proton energies for each target and monitor foil were determined by range energy calculation using a computer program originally based on the polynomial approximation of Ziegler [cf. 27]. The energy scale was checked additionally by comparing the measured excitation function of the three monitor reactions with their recommended values.

Determination of the beam energy of a variable energy multiparticle cyclotron of type MGC-20 is required for accurate measurements of reaction cross sections, and optimization of commercial production of GBq batches of radioisotopes by using the threshold nuclear reactions, which provide us with a precise knowledge of the particle beam energy. Literature study of recently published articles on cross-section determination shows a number of reactions with well-established values for their cross section behavior. Experimental determination of the cross sections for these reactions often referred to as monitor reactions, and comparing them with the literature values allows an estimation of the incident beam energy. The nuclear reactions of interest in beam monitorng are $^{nat}Cu(p,x)^{62}Zn$ [Kopecky, (1985); $^{nat}Ni(p,x)^{55}Co$ (Tarkanyi et al., 1991); $^{nat}Cu(p,x)^{65}Zn$ (Kopecky, 1985); $^{nat}Ni(p,x)^{57}Ni$ (Tarkanyi et al., 1991) [cf. **95**-**98**].

Target stacks of 10-15 foils (99.9 % pure) were irradiated for 60 min with external beam currents up to ~0.2 μA. Yields and cross sections of the nuclides of interest were calculated from the γ emission rates using disintegration data from Firestone et al., (1996) [cf. 4].

Figure (3-4) shows the agreement between the recommended cross section variation for monitor reactions and the corresponding experimental data of this work are plotted as a function of the mean energy in each foil, with proton energy from thresholds up to ~20 MeV for $^{nat}Cu(p,x)^{63}Zn$, $^{nat}Cu(p,x)^{65}Zn$ reactions induced in Cu foils produce ^{63}Zn, ^{65}Zn, and $^{nat}Ti(p,x)^{48}V$ reactions induced in Ti foils produce ^{48}V, respectively. The candidate incident energy $E_{incident} = E_b^{cal}$ calculated by using cyclotron parameters, and the other incident mean energy derived from Eq. (3-1) in each foil, obtained applying the activation technique $E_{incident} = E_b^{act}$. The degradation in beam energies are computed from Ziegler's polynomial estimation formulae [cf. **27**], After applying any proposed energy shifts deduced from activation technique, we obtain much better agreement with the reference values for the cross-sections.

$$E_m = \frac{\int_0^{E_1} E\cdot\sigma(E)\cdot dE}{\int_0^{E_1} \sigma(E)\cdot dE} \qquad (3-1)$$

Where σ(E) is the cross section at the energy E and E_1 is the incident energy. In Eq. (3-2) instead of using cross section values one can also use thick yields, which are easier to measure at different energies (Ishii, 1978) [cf. **99**], the mean beam energy formula becomes:.

$$E_m = \frac{E_1}{1 - \left[E_1 \Big/ A_0(E_1) \right] \int_0^{E_1} \left[A_0(E) \Big/ E^2 \right] \cdot dE} \qquad (3-2)$$

Where, $A_0(E)$ = thick target yield in the given target at energy E.

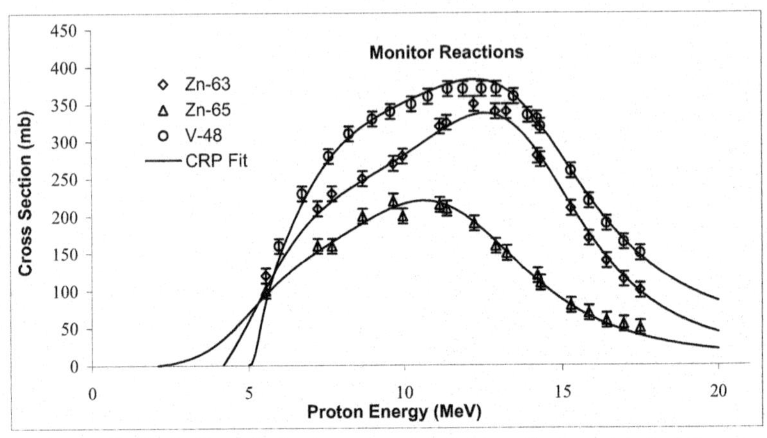

Figure (3-4): Indicates the Agreement between Recommended Cross Section Variation for Monitor Reactions and the Corresponding Experimental Data of this Work.

It is possible to apply the average stopping power method also for a "defined thickness" target. In this case, Eq. (3-3) becomes:

$$E_m = \frac{\int_{E_2}^{E_1} E \cdot \sigma(E) \cdot dE}{\int_{E_2}^{E_1} \sigma(E) \cdot dE} \qquad (3-3)$$

Following the same procedure as for a thick target, we obtain for a "defined thickness" target as:

$$E_m = \frac{A_0(E_1) - A_0(E_2)}{\left[A_0(E_1) \Big/ E_1 \right] - \left[A_0(E_2) \Big/ E_2 \right] + \int_{E_2}^{E_1} \left[A_0(E) \Big/ E^2 \right] \cdot dE} \qquad (3-4)$$

When E_2=0, Eq.(3-4) reduces to Eq.(3-2).

With known incident energy E_1 the energy E_2 of the particle that leaves the target can be calculated theoretically by knowing its thickness, from the range energy tables or more precisely by measuring the activity ratio of two nuclides from different nuclear reactions induced in a suitable monitor placed in front or behind the target. Such an energy measurement which is independent of foil thickness, flux etc., is relatively simple and less erroneous than an absolute method.

In some cases, for nuclear reactions with high threshold values, the thickness of the sample can be so "defined" as to give E_2 a value below the threshold for the nuclear reaction with the sought element. So that any small error in measuring E_2 will have a negligible influence on the calculation of E_m. The uncertainty in the incident energy estimation derived from excitation curve matching depends furthermore on the slope of the cross-section plots in presence of energy shift.

3-5 Beam Intensity Monitoring

One of the most important parameters in the determination of a cross section is the acculate measurement of the beam current incident on each foil. The beam current is measured using a Faraday cup as an indicative value. This value should be corrected with the values determined from the measured activities of the foils adjacent and surrounding the specified sample under investigation. These foils are called monitor foils. Both methods are employed simultaneously to achieve more accuracy. The actual value of the beam current during irradiation is estimated from the measured activity of the monitor foils using the activation formula Eq.(1-23), where the recommended cross section is substituted from the tabulated data of the monitor reaction.

The choice of the monitor material depends on several considerations, such as:

(1) Monitor should have a high cross section within the energy range of irradiation.

(2) The monitor reaction cross sections used should have accurate values, recommended, and up to date.

(3) The monitor products should have compatible half-lives with the resulting nuclides.

The beam current value is a very significant parameter in the calculation of the cross section. It depends mainly on the uncertainty of the monitor reaction cross section. Also it can be affected by the geometrical scattering of the beam within the stack. The uncertainty of the measured activity arises mainly from the uncertainty in the efficiency of the counting system. In most cases the efficiency is known well so that in general the total uncertainty due to this source does not exceed 5%. The uncertainties of energy degradation within the stack lead to shift in the energy scale up to ±1 MeV, depending on the thickness of the stack and the homogeneity of the irradiated samples. The cross section data reported in literature have general uncertainties of about 10 to 25%.

The irradiations were carried out in the target holder at the end of a beam line, in two geometeries external, and internal irradiation, external which, at the unit which is used in the radionuclide production. Internal where, the sample irradiated inside the beam guideline in vacuum, the beam hit the target normally. So, both are equipped with cooling system, a long collimator, electron suppresser, and current measurement. At 18 MeV nominal primaries beam energy, and in each case the beam was stopped in a thick Al disc immediately behind the stack. The irradiation was performed with a 10-mm diameter at ~200 nA intensity for one hour. The beam intensity was kept constant during irradiation. The beam current measured with Faraday-Cup was corrected after analyzing the results of the monitor reactions. High purity foils of Cu, and Ti served as simultaneously beam monitor foils for the accurate measurement of the incident proton flux. For this purpose, we used high purity natural Cu foils of 10μm thick (99.9%), and high purity natural Ti foils of 25μm thick metallic natural foils (99.9%). The thicknesses of the stacks were sufficient to provide adequate energy overlaps of the successive data sets for the consistency of the data to be thoroughly checked.

3-6 Associated Counting Electronics

The activation products were identified by the characteristic energies of γ-ray lines and their half-lives. The γ-activities in the samples were recorded using high performance gamma ray spectrometer.

3-6.1 γ-Spectrometer

Proton-induced activities were determined using the counting system consisting of vertical coaxial closed end 70% HPGe detector (Canberra, 2 keV full width at half-maximum at energy of 1332.5 keV), detector bias supply (Ortec-447), main amplifier (Ortec-570 amplifier), multichannel analyzer buffers, with built-in ADC digitized the pulses into spectra of 8192 channels (TRUMP-8192), and lead shield of thickness 10cm surrounds the system.. The connection of the components of the complete gamma ray spectrometer system is shown in Figure (3-5).

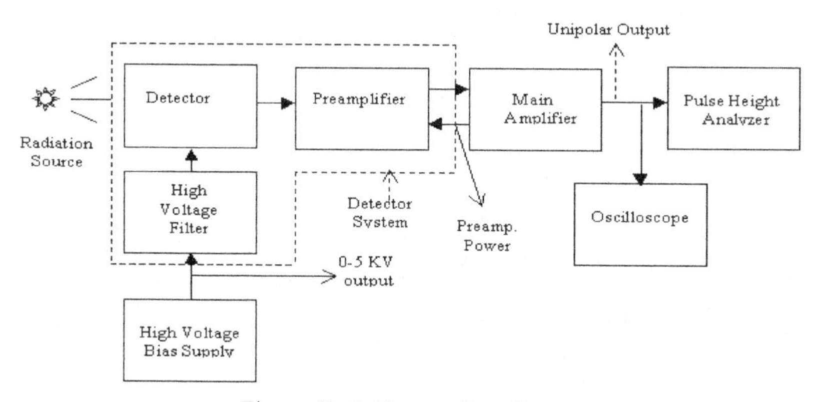

Figure (3-5) Gamma Ray Spectrometer

The γ-spectrometric measurements started after the end of irradiation. The amplification was chosen for the registration of γ-quanta with energies between some tens of keV and about 2MeV. Typical resolutions ranged from about 1 keV at 122 keV of ^{57}Co to about 2 keV at 1332 keV of ^{60}Co. The absolute calibration of the γ-spectrometer was performed for each geometry used by calibrated radionuclide sources (^{22}Na, ^{57}Co, ^{60}Co, ^{137}Cs, ^{133}Ba, $^{152,154,155}Eu$, ^{109}Cd, ^{65}Zn, ^{54}Mn, and ^{232}Ra) as seen in table (3-3) with certified accuracies of ≤2% (Oak Ridge, Tennessee, USA).

To avoid problems with high dead-times and pile-up effects the distances between sample and detector window were varied between 5 and 50 cm. Thus it was possible to keep the dead time below 10% with no detectable pile-up effects. The large distances were used especially for the short $T_{1/2}$ measurements shortly after the irradiations because of the high activities of the samples. In these geometries no lead shielding of the detector was possible but interference with background γ-lines were negligible anyway due to the short counting times as well as due to the high Compton-background caused by the measured samples themselves.

For the latter measurements we used lead-shielded detector with shield of 10cm thickness, having an additional 5mm low activity copper shielding inside. Background spectra were taken for these measurements. These spectra were used in the data evaluation procedure to correct measured activities for background interference. Measurement times ranged from 10 min., at the beginning up to about 2 hours in some cases at the end of a measurement series.

The spectra of the irradiated samples were analyzed, and the gamma lines of produced radionuclides were identified. Figures (3-5), (3-6) represent samples of the recorded spectra of the irradiated monitors of ^{nat}Cu, and ^{na}Ti. The areas under the peaks of the gamma lines 596.6, 669.6, 1115.5, and 983.5 keV as illustrated in these figures were used for the calculation of the reaction cross section of the radionuclides ^{62}Zn, ^{63}Zn, ^{65}Zn, and ^{48}V, respectively. The gamma lines were well separeated and no gamma lines overlaps were observed.

3-6.2 HPGe-Detector Efficiency

In the case of measuring proton fluence, we usually do not use point like samples but extended samples, this causes a common problem in the quantitative analysis of these samples by gamma spectrometer, see Figure (3-7). If we seek to achieve absolute result of high accuracy, then the gamma-ray detection efficiency for geometrically extended sources must be well known. It is therefore, important to have available method to correct measurements taken with HPGe for the effect of finite extended samples, bearing in mind that the HPGe detector efficiency curves that are most often and conveniently determined using point-like standard sources.

So to perform that, the HPGe detector efficiency was first determined using a variety of point radionuclides. Using different gamma ray emitters placed on the axis of the detector at several distances from its face, the full energy peak efficiency (FEPE) of the detector from 60 to ~2000 keV gamma-ray energies was achieved with 2.2% overall inaccuracy. Then we used Ganaas program [cf. **100**] in estimating the correction coefficients for extended samples (disc) by correcting the reference efficiency data for detector-source solid angle, gamma attenuation through source-detector space, gamma self absorption in the samples, and random and true coincidence summing corrections for these extended samples.

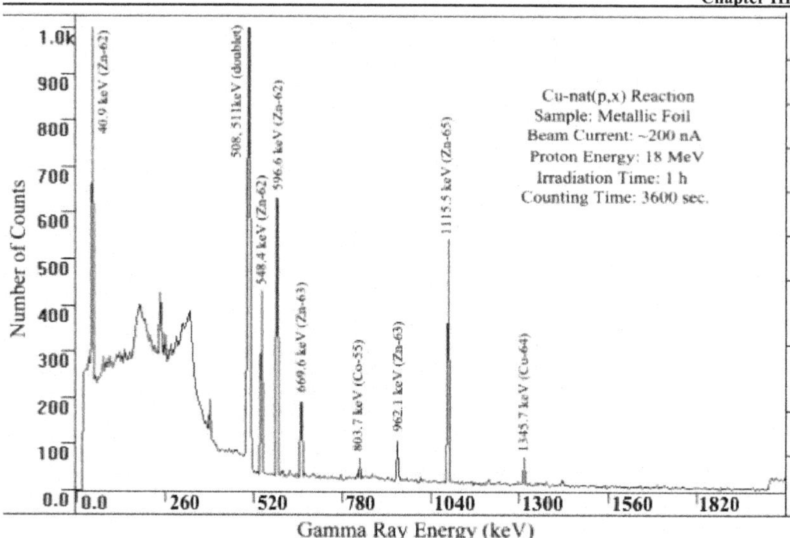

Figure (3-6): Sample of Gamma Ray Spectrum of Natural Copper Sample Irradiated with Protons for Beam Intensity Monitoring.

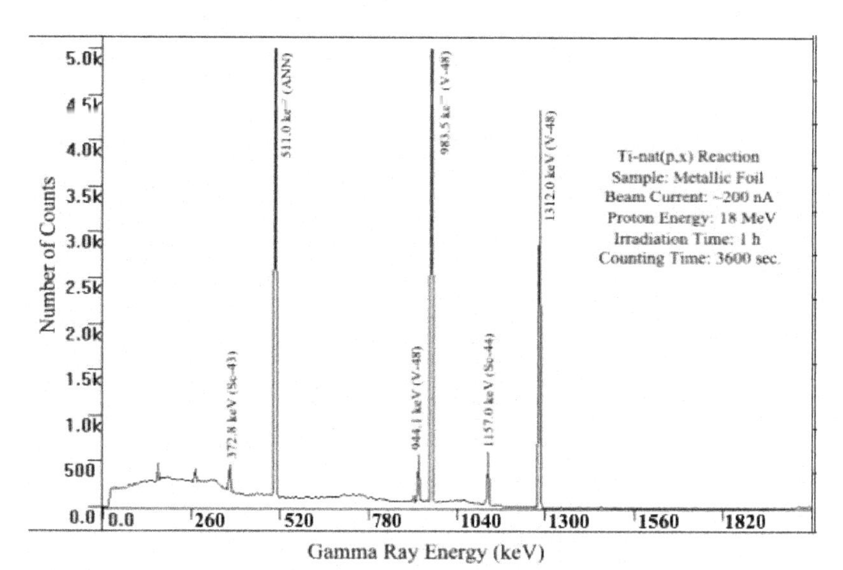

Figure (3-7): Sample of Gamma Ray Spectrum of Natural Titanium Sample Irradiated with Protons for Beam Intensity Monitoring.

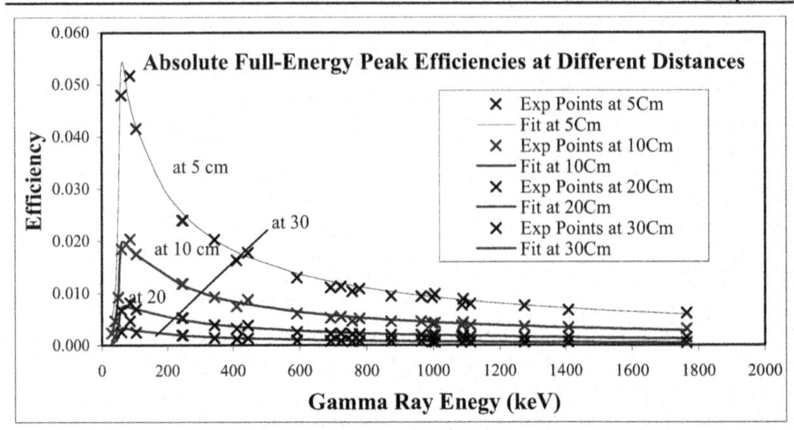

Figure (3-8): Absolute Full Energy Peak Efficiency of HPGe-Detector, Experimental Points, and Fitted Curves at Different Distances.

The gamma ray points-like standard sources used for the experimental determination detector efficiency are listed in Table (3-3). A typical plot for the full energy peak efficiency of HPGe-detector with source to detector distances 5, 10, 20 and 30 cm is shown in figure (3-8). These sources cover gamma ray energies from 60–2000 keV. The standard sources were positioned on axis above the cap of the detector at distances ranged from 5 to 50cm from detector end cap by placing them on lightweight card tray atop light weight card cylinders. Distances were measured from the center of the source to the external front face of the detector end cap. Following the corrections for dead time and room background, the number of counts C in the full energy peak (FEP) was obtained by fitting a Gaussian profile superimposed on a linear background to the spectrum. This was related to the absolute efficiency ξ in count per gamma emitted according to [cf. **91**]:

$$\xi\left(E_\gamma\right)=\frac{C\left(E_\gamma\right)}{A_o\cdot I_\gamma\cdot\exp\left(-\lambda\cdot t\right)} \qquad (3-5)$$

where:

$C(E_\gamma)$: the count rate of the peak corresponding to a certain γ-ray energy E_γ;

A_o: the activity of the sample (Bq) at some arbitrary reference time;

I_γ : the absolute γ-intensity;

λ : the decay constant.;

t : the decay time.

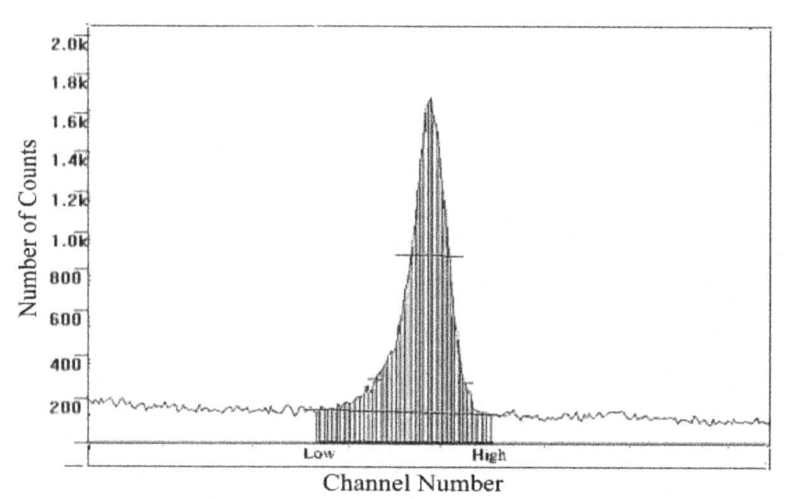

Figure (3-9): Evaluation of one Spectrum Peak by Using Gaussian Fit, and Linear Background Subtraction

Table (3-4): Gamma Ray Standard Sources Used in the Efficiency Determination.

Standard Source*	$T_{1/2}$	Standard Source*	$T_{1/2}$
^{133}Ba	10.52	^{154}Eu	3140 y
^{109}Cd	462.2 d	^{155}Eu	1740 y
^{57}Co	271.79 d	^{54}Mn	312.3 d
^{60}Co	5.271 y	^{22}Na	2.602 y
^{137}Cs	30.19 y	^{226}Ra	620500 y
^{152}Eu	4960 y	^{65}Zn	244.26 d

*Supplied from (Oak Ridge, U.S.A.).

Figure (3-9) shows an evaluation of one spectrum peak by using gaussian tail fit, linear background subtraction. A six-order logarithmic energy function was used to fit the plot. The efficiency functions were obtained by fitting a logarithmic energy negative power transfer series function to the experimental efficiency point data from 60 keV up to 2 MeV. The efficiencies were repeatedly checked for every-geometry. Energy calibration was done using a least-squares fit to a third-order polynomial for the energies of the calibrated radionuclide sources.

Table (3-5): Parameters Used to Fit the Efficiency Function Eq. (3-6).

Par.	Fit parameters Value at Different Distanced			
	At 5cm	At 10cm	at 20cm	at 30cm
A_0	-5.60497	-3.60156	-1.75206	-0.58176
A_1	-3.74942	-3.55185	-3.42532	-3.35633
A_2	-1.66771	-1.56893	-1.50566	-1.47117
A_3	0.195127	0.260982	0.30316	0.326156
A_4	0.423646	0.473037	0.50467	0.521917
A_5	0.529916	0.569429	0.594736	0.608534

The fitting function used in efficiency calculation at 5,10,20, and 30cm is the logarithmic energy power transferred series, it is defined as:

$$\varepsilon(E) = \left(1/E\right) \cdot \sum_{i=0}^{5} \left(A_i \cdot \ln(E)^{-i}\right) \qquad (3-6)$$

Where, E is the gamma ray energy; A_i is the fitting parameters. Table (3-5) shows these parameters (A_i) of the experimental points using the above-mentioned function obtained from running the Monte Carlo program [cf. **100**].

3-7 Analysis of Gamma Spectra

The analysis of the γ-spectra, the identification of nuclides and the selection of activities used in the final calculation of cross-section are being the most time consuming step in the experiments. Figures (3-10):(3-13) represent the γ-spectra for some samples of the irradiated targets for ^{nat}Mo, and ^{nat}Gd. The γ-spectra are not extremely complex, exhibiting quiet separated peaks. The gamma lines of energies 934.4, 702.5, 204.1, 765.8, 778.2, 140.5, 344.3, 540.1, 996.3, 262.3, 534.3, and 879.4 keV are illustrated in these figures. These gamma lines were used for the calculation of the reaction cross section of the radionuclides ^{92m}Nb, ^{94}Tc, ^{95m}Tc, ^{95g}Tc, $^{96m+g}$Tc, ^{99m}Tc, $^{152m+g}$Tb, ^{154m}Tb, ^{154g}Tb, ^{155}Tb, ^{156}Tb, and ^{160}Tb, respectively.

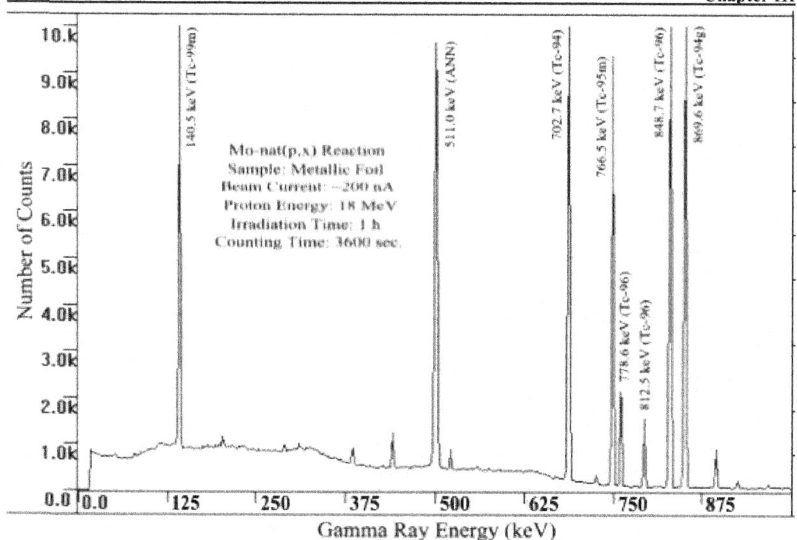

Figure (3-10): Samples of Gamma Ray Spectra of Natural Molybdenum Sample Irradiated with Protons for Reaction Cross Section Measurements one hour after EOB.

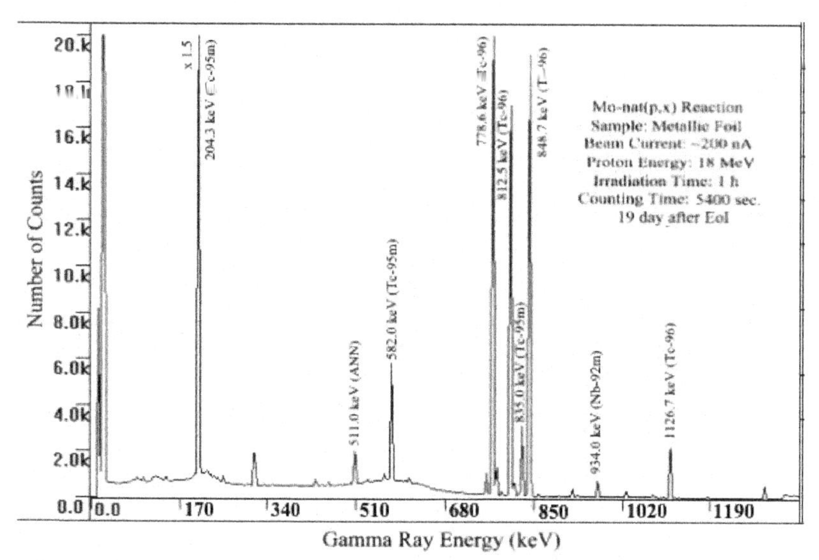

Figure (3-11): Samples of Gamma Ray Spectra of Natural Molybdenum Sample Irradiated with Protons for Reaction Cross Section Measurements 19 day after EOB.

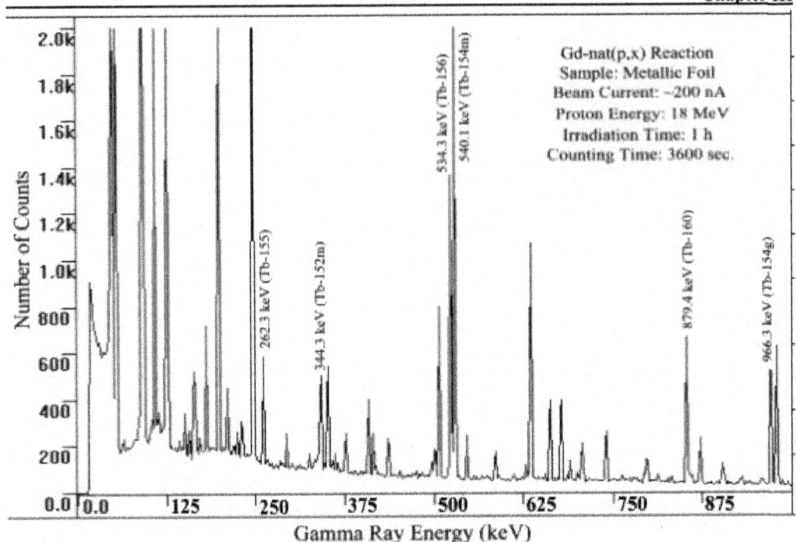

Figure (3-12): Samples of Gamma Ray Spectra of Natural Gadolinium Sample Irradiated with Protons for Reaction Cross Section Measurements one hour after EOB.

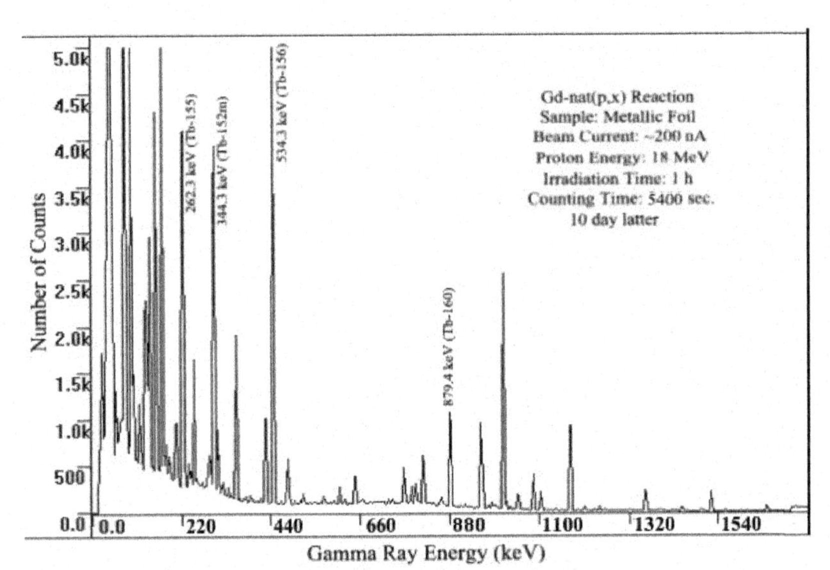

Figure (3-13): Samples of Gamma Ray Spectra of Natural Gadolinium Sample Irradiated with Protons for Reaction Cross Section Measurements 10 day after EOB.

The complexity of the spectra has two reasons. Firstly, due to adjacent of some peaks close to each other, the net counts of some peaks depend on the presence of the adjacent peaks. Secondly, dues to hereditary origin or pile up contribution to the continuum background. After short decay times the spectra are dominated with some long-lived radionuclides which are produced with large cross-sections also have quiet separate peaks.

The analysis of the γ-spectra was done using the commercially available code APTEC-MCA [cf. **101**]. APTEC-MCA to calculate the net peak areas via an unfolding algorithm using a least-squares fit [cf. **102**,**103**]. For an evaluation, regions of a spectrum are defined in which all peaks are unfolded simultaneously after the background has been calculated. Peak shapes are assumed to be Gaussian with a low-energy tailing. For each detector and measuring geometry, parameters were determined and supplied to the code which is a function of energy or channel numbers the full-width at half-maximum of full-energy peaks and their tailings [cf. **103**,**104**].

The user can do the whole evaluation procedure either in an automatic mode or interactively. Although APTEC-MCA is a sophisticated and successful code, detailed tests showed in our case that the automatic mode is not reliable enough with respect to the necessary region of the spectrum, peak recognition, background determination, and multiplet deconvolution and net-peak area calculation. Thus, we analyzed each spectrum interactively, making sure to get a maximum of information out of each spectrum. Proceeding in this way means, however, to give up the reproducibility of a spectrum analysis in contrast to the automatic and strict application of a mathematical algorithm. Therefore, repeated spectra analyses were performed independently by both modes. The results obtained in this way always were in good agreement within the limits of uncertainties of the unfolding procedure.

3-7.1 Nuclide Identification

Besides the spectrum analysis, an automatic and unambiguous identification of product nuclides turned out to be rather complicated. Most codes and also APTEC-MCA offer such capabilities, with a general applicable algorithm, which can be used, with some confidence for complex spectra. But capable to assign the isotope library file which concern the region of isotopes, match the ROI centroide energy with the energy of isotope within a tolerance of plus minus fraction of peak in keV or its full width at half maximum. Fraction limit value, before accepting a peak as a possible match to a certain isotope check the active library to determine where any or all other lines for that isotope would be located. These lines are then checked to see if there is a ROI (i.e. peak) at that location. If there is a match at the other energies, then the yield at that ROI is compared to the yield at the line in question. The identification process starts with the spectra having the longest decay times. Then, successively going back in time, the identified nuclides are searched in the preceding spectra by calculating their respective activities and looking whether these activities should be detectable in the spectrum under consideration. Applying this algorithm decreases the number of possible nuclides in a spectrum. Table (3-5) contains the formulae repeatedly used in the procedures for bumerical analysis of spectra's peaks.

The identification of radionuclides was only performed for γ-energies above 60 keV. Below this energy it was impossible to get quantitative results due to the variety of overlapping x-rays and γ-rays of the produced radionuclides. In the step of identification of nuclides, the half-lives from the Table of Isotopes Firestone et al., (1996) [cf. **4**], energies and branching ratios of γ-rays were taken too. Samples of the decay schemes of the produced radionuclides obtained in this study are presented in the figures (3-14) to (3-28).

Table (3-6): Formulae Used for Numerical Analysis of Spectra.

Quantity	Formula
$Peak\ Area\ (A)$	$\sum(C_x - B_x)$
$Peak\ Statistical\ Precision$ $\%Error\ \delta\bar{A}\ \ (1\,Sigma)$	$\frac{\sqrt{\sum(C_x + B_x)}^{*}}{A}\cdot 100$
$Centroid\ Weighting\ (\bar{x})$	$\frac{\sum(C_x - B_x)\cdot x}{A}$
$Centroid\ Precision$ $\delta\bar{x}$	$\frac{\sqrt{\sum(C_x + B_x)\cdot(x-\bar{x})}^{*}}{A}$
$Half-Width\ (\bar{w})$	$\sqrt{\frac{\sum\left[(C_x - B_x)\cdot(x-\bar{x})^2\right]}{A}}$
$Half-Width\ Pricision$ $\delta\bar{w}$	$\frac{1}{2\bar{w}A}\sqrt{\sum\left[(x-\bar{x})^2\cdot\left(1-2\cdot\frac{C_x - B_x}{A}\right)-(\bar{w})^2\right]^2\cdot[C_x - B_x]}^{*}$

$Where,\ (C_x) = Counts\ in\ channel\ x,\ (B_x) = Background\ in\ channel\ x$

$*\ Error\ formulae\ based\ on\ the\ conservative\ assunption\ that\ \delta B_i = \sqrt{B_i}$.

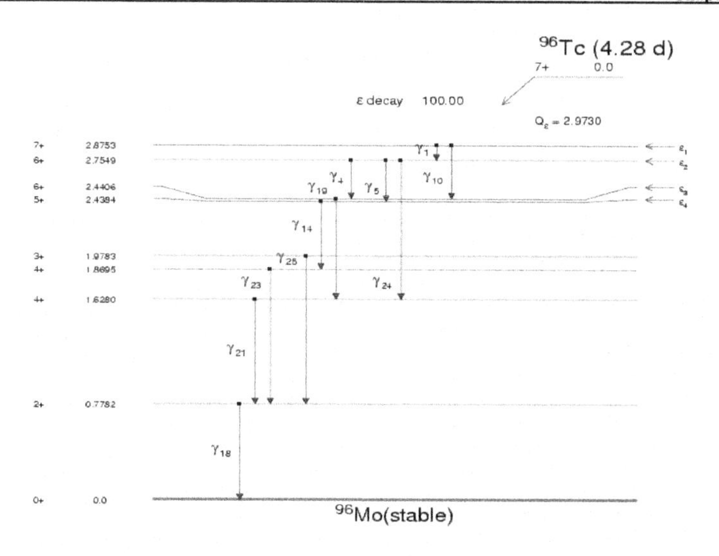

Figure (3-14): The Decay Scheme of ^{96}Tc [cf. **4**].

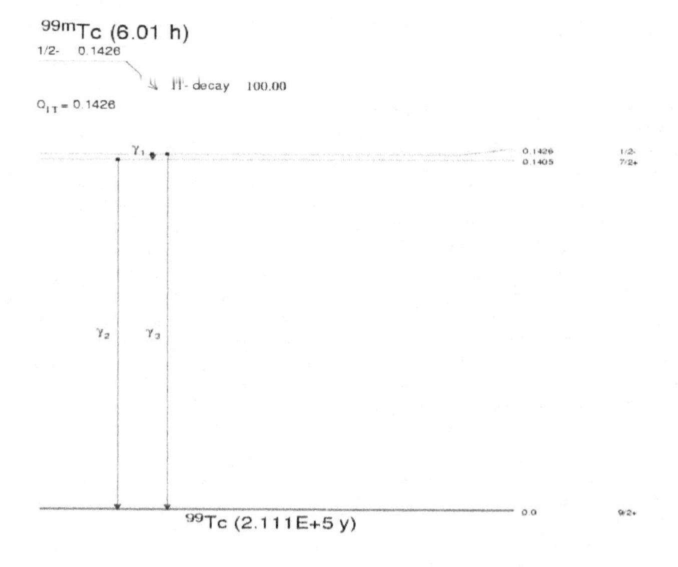

Figure (3-15): The Decay Scheme of ^{99m}Tc [cf. **4**].

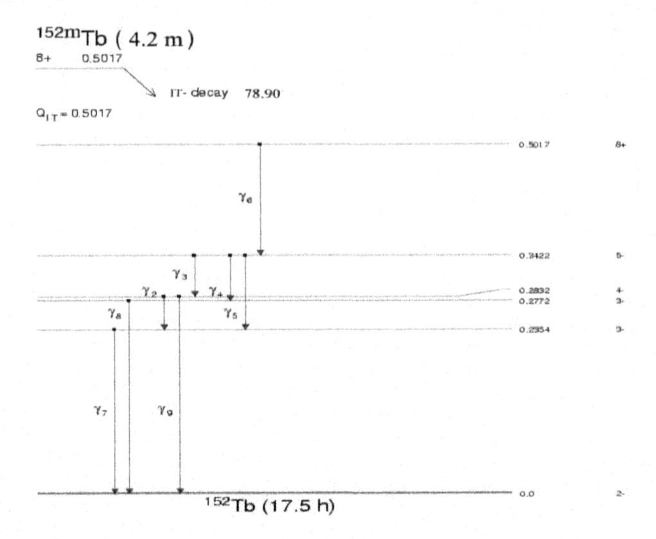

Figure (3-16): The Decay Scheme of ^{152}Tb [cf. 4].

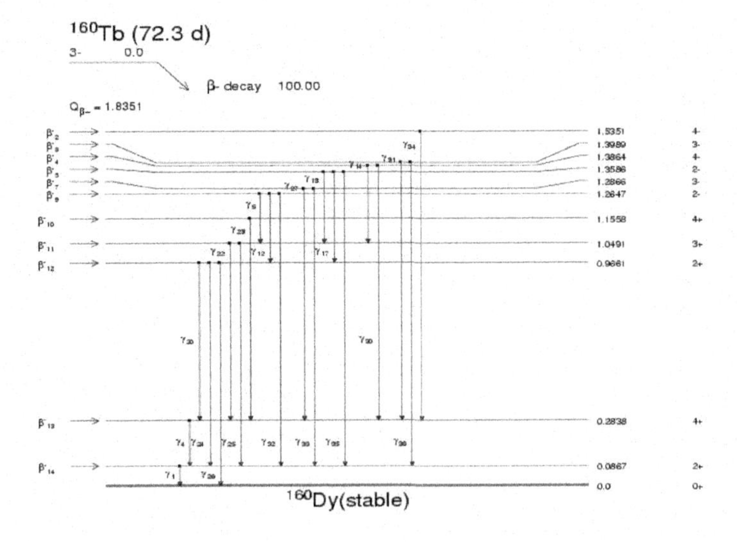

Figure (3-17): The Decay Scheme of ^{160}Tb [cf. 4].

3-7.2 Selection of Reliable Data

After nuclide identification and calculation of their activities for each target, the last step before calculating cross-sections is to gather all activities calculated from the spectra in one file sorted by nuclide. By this way we get a survey over all measurements for each nuclide and its γ-lines allowing for half-life control as well as checking for consistency of activities determined using different γ-lines of a nuclide. By selecting activities, which are judged to be reliable, cross-sections are calculated as described below.

This judgment is performed on the basis of objective and subjective criteria. Objective criteria are, e.g. the agreement of the activities derived from different γ-lines of a nuclide and constancy of the activity calculated for the end of irradiation from several measurements (our half-life control). Subjective criteria are based on the experience and knowledge of the evaluator about the entire spectrometric method and their characteristic properties, about spectrometric interference and about the physics of the underlying problem.

3-7.3 Definition of the Detection Probability

Suppose that subsequent to an induced reaction, the radioactive decay of the produced nucleus takes place via a given cascade C_J consisting of a number of individual γ-rays of energy E_1, E_2, …, E_i. The probability $W_j^{(t)}$ of detecting by a HPGe detector counting a reaction event using a target of thickness δ can be expressed as a product of six functions, i.e.,

$$W_j^{(t)} = \sum_i P_j(E_i) G_j(\Omega, E_1, \dots, E_i) S_{ai}(E_i, \delta) C_{Si}(E_i, E_j) D_t(C_g, \tau) F(E_i, t). \qquad (3-7)$$

Which represent the total sum for correcting the erroneous estimates of uncertainty magnitudes concerning some error sources. Where:

$P_j(E_i)$ is the probability that a gamma ray of energy E_i belonging to the cascade C_j will be emitted. $P_j(E_i)$ will be referred to as the "emission probability" due to the probability of γ-ray belonging to competitive reaction channel, for low gamma or x-ray that repeatedly suffer from electron-capture fraction, internal conversion coefficients, and fluorescence yields.

$G_j(\Omega, E_1, \dots, E_i)$ the probability that a gamma ray of energy, belonging to the cascade C_j and unaccompanied by other coincident gammas will enter the detector which subtends an effective solid angle, Ω, at the source. This will be referred to as the "geometrical probability".

S_{ai} is the probability that photons emitted within a differential geometry solid angle $d\Omega$ and impinging on the detector, interact incoherently with the active detector crystal. Including the attenuation factors through the source itself whenever by using non-selfabsorbence standard source and samples also due to Air gab, end cap cover, Aluminum or Beryllium window, fronted dead layer of detector, and any absorber introduced between detector source geometry.

C_{si} true coincidence correction that involved in measurements of a radionuclide that emits cascade radiation. When two or more cascading photons give arise to a total or partial energy deposition in the detector within the response (or resolving) time of the detecting systems, coincidence detection of these photons appears. True coincidence may result in summing effects or loss effects in the forms of $\gamma-\gamma$, $\gamma-x$ ray due to internal conversion (IC), $\gamma-$ x ray due to electron capture (EC), etc., [cf. **105**,**106**]. The significance of these effect increases with increasing counting efficiency, therefore, the correction for true-coincidence effects is of vital importance.

D_t accidental coincidence summing, their number could be measure experimentally by a recently developed a Multi-channel time scaling technique, but it is usually obtained by means of appropriate mathematical formulae which depend on the value of the dead time used in the experiments [cf. **107**,**108**]. The accidental coincidence on the vice of true-coincidence which is independent of resolving times, dead times, and baseline fluctuation. The pile up effects especially in high count rate spectroscopic circuits that cause the distortion of pulse height spectra. A convenient computation used to analyze the distortion of the spectra, taking into account the analyzer dead time. The distortion of the spectra can be described as follow. Pulses from a detector having a finite duration are statistically distributed in time and their superposition generates a random process at the amplifier output, which can be considered partially as a baseline fluctuation at the analyzer input.

$F(E_i,t)$ the probability that a γ-ray of energy; belonging to the cascade C_J and originated from produced radionuclide due to reactions induced by primary proton fluence, secondary build up particle, gamma fluence may induce production of the same residual radionuclide or others. This in course may alter what so-called "effective primary beam". That causes an increase or decrease on the primary beam and also tempt a degrade on the primary proton fluence and it acts due to secondary beam fluence through the stacked foils, which will yield an effective fluence pulse. $F(E_i,t)$ is termed as the "Fluence Response Probability".

3-8 The Sources of Uncertainties

The following major sources of uncertainties were considered and taken into account according to the laws of error propagation, in the total errors quoted for the measured cross sections. For the results obtained two types of uncertainties have to be considered: those of the proton-energies in the targets and those of the cross-sections themselves. There are many sources of uncertainties of the cross-sections and some may result in considerable uncertainties. The following sources of uncertainties were considered.

3-8.1 Uncertainties in Proton Energies, Beam Fluctuation

For calculating the uncertainty of the proton energy in a target, three factors have to be considered. First, the protons leaving the accelerator have an energy uncertainty ΔE_A. Second, the protons are slowed down in the target foil from an initial energy $E_{n,i}$ to the final energy $E_{n,f}$ (resulting in an energy spread with a half-width $\Delta E_{loss} = (E_{n,i} - E_{n,f})/2$. Third, due to the statistical nature of the slowing down process there is an energy straggling which can be described by a Gaussian distribution with a width (straggling) parameter α_i. The uncertainties in range-energy calculations lead to errors in the energy scale. Use of thick foils in the case of low energy projectile leads to considerable errors. Especially, in the case of sharply increasing or decreasing cross-section of an excitation function. Consequently, the uncertainty of proton energy E_n in the nth target foil of a stack over all three-type losses is given by:

$$\Delta E_n = \sqrt{\Delta E_B^2 + \Delta E_{loss}^2 + \left(\sum_{i=1}^{2} \alpha_i\right)}. \qquad (3-8)$$

The fluctuation in beam intensity during irradiation is esenssial, the constancy of flux density over irradiation time could not be ensured. The beam intensities were continuously monitored and recorded. The fluctuations and interruptions of the irradiations were taken into account when calculating fluxes and cross sections. Using the following replacement in the activation formula Eq. (1-23) Gloris et al., (2001) [cf. **109**].

$$\frac{1}{1-\exp\left(-\lambda t_{irr}\right)} \to \sum_{i=1}^{n} \frac{\exp\left(\lambda\left(t_{EOI}-t_{EOI,i}\right)\right)}{1-\exp\left(-\lambda t_{irr}\right)} \qquad (3-9)$$

Because the flux density itself may not be constant during irradiation we replaced Φ in Eq. (1-23) by:

$$\Phi \to \frac{1}{t_{EOI}-t_{BOI}} \int_{t_{BOI}}^{t_{EOI}} dt^{*}\, \Phi\left(t^{*}\right). \qquad (3-10)$$

With Φ(t*) being the relative measurements of the beam current. Using these replacements the uncertainties due to fluctuations in the beam intensity become negligible. Moreover, they would affect only the cross-sections for very short-lived nuclides.

3-8.2 Uncertainties in Net Peak Area, Deacy Data

The uncertainty of each single net-peak area is determined by the spectrum evaluation code. It takes into account the Poisson uncertainties of the counts in the individual channels. As well as the uncertainty of the background determination, propagating them according to the law of error propagation through the unfolding procedure. Implying reproducibility of photopeak integration, and Statistical counting errors. Sometimes a peak cannot be attributed unambiguously to a single nuclide. If the contributions were not negligible and the activity of one of the contributing nuclides can be determined using another line or in a later spectrum, the interfering lines were corrected using this activity. If it seemed that the contributions of other nuclides to a peak are very small, no correction was applied. Due to this procedure we assume a maximum inaccuracy of 2% due to contributions of other nuclides but it must be pointed out that on the average this uncertainty should be smaller.

The half-lives of produced nuclei were taken from Firestone et al., (1996) [cf. 4]. Larger uncertainties in any time dependent terms in calculation of the cross-sections via Eq. (1-23) would result in observable disagreements between different measurements due to the exponential dependence on time. Uncertainties of the half-lives were considered to be of the order of 1 % for the nuclides measured. Larger uncertainties would have shown up during the half-life control.

The γ-abundances of residual nuclei were also taken from Firestone et al., (1996) [cf. 4]. Since we used normally the strongest γ-lines, which were rather well known, we assume a global uncertainty of 1%. However, it must be pointed out that the errors given for the cross sections also account for inconsistencies between the absolute γ-intensities of the γ-lines used.

3-8.3 Uncertainties in the Absolute Efficiency, Time Factor, and Dead Time

Uncertainty of the absolute efficiency calibration of γ-spectrometers could be deduced from the comparison of different calibrated radionuclide sets, The calibration standards used had a certified accuracy of ~2%. The error of absolute efficiency calibration can be adopted with a total uncertainty of the full energy-peak efficiency of 4.5%. It is larger than the error of relative efficiency calibration. From comparisons of measurements performed on different detectors, the relative uncertainty of the efficiency calibrations can be considered to be 2%. Since the error in the absolute efficiency calibration affects the flux determination.

Uncertainty of time factor represents the uncertainties in the irradiation time, decay time and counting time: For time scales determined by the half-lives of nuclides observed within this work, we assume that uncertainties of the above-mentioned quantities are negligible.

Dead time and pile-up losses in the γ-spectrometer were automatically corrected. Pile-up effects were not seen because the distances between samples and detector were varied in the way that the counting rates were low enough (kept below 10%) to avoid both pile-up and failure of the automatic dead-time correction. Where, the internal life times correction of the ADC/MCA system was checked to be correct.

3-8.4 Uncertainties in Irradiated Nuclei, Impurities, Recoil Contaminations

The number of irradiated nuclei in each foil was well known. Where, each foil was weighted with an absolute uncertainty of ±30 μg. In case of the Gadolinium foils of 10 μm equivalent to 7.901 mg cm^{-2}, this resulted in an inaccuracy of about 1%, with pinhole free and good homogeneity. While the uncertainty for the heavier foils, with weights of about 20 or 25 μg becomes negligible.

Uncertainty of impurities in the target materials causing interfering nuclear reactions and errors resulting from the correction of contributions of other target constituents: Due to the high purity of the target foils used, contributions of other constituents were not considered.

Recoil losses, and recoil contaminations could be assigned by analyzing the foils in the inner part of each "target element unit" consisting of more than two identical foils (mini-stack) these errors cancel out. For target elements from vanadium to cobalt the mean ranges of recoil nuclei R can be approximately calculated by the empirical formula R $(mg/cm^2) \approx 0.045 \cdot \Delta A$ [cf. **110**] with ΔA being the difference between target and product masses. The recoil contamination from the adjacent foils had to be corrected.

3-8.5 Uncertainties due to γ-Interference γ-Self Absorption, and γ-Attenuation

From the very complex spectra, the cross-sections for some nuclides could only be determined after the correction of interfering γ-lines from another nuclide, which could not be resolved by our detectors. Assume that $A_1(t_{EOB})$ and $A_2(t_{EOB})$ are the activities of two nuclides at the end of irradiation which have interfering γ-lines with $I_{\gamma 1}$ and $I_{\gamma 2}$, being the abundance of the corresponding γ-quanta. Then first a wrong activity $A_1^*(t_{EOB})$ is calculated according to Eq. (1-23) in our evaluation procedure under the assumption that the peak is only caused by nuclide 1. If $A_2(t_{EOB})$ is known, e.g. from other γ-lines of nuclide 2, we can calculate the correct value of $A_1(t_{EOB})$ to be [cf. **22**]:

$$A_1(t_{EoI}) = A_1^*(t_{EoI}) - A_2(t_{EoI}) \cdot \frac{\lambda_1 I_{\gamma 1}}{\lambda_2 I_{\gamma 2}} \cdot \frac{1-\exp(-\lambda_2 t_C)}{1-\exp(-\lambda_1 t_C)}$$
$$\times \exp(-(\lambda_2 - \lambda_1) t_d) \qquad (3-11)$$

Although it is in principle possible to apply this scheme to more than only one interfering γ-line, we limit ourselves to one correction term since the resulting uncertainty of the corrected activity quickly becomes rather high if the correction is large.

Self-absorption of γ-rays in the sample, and in the aluminum catcher foils played no role, it did in the sample foils. For γ-energies above 120 keV and for thin, light- and medium weight-target elements this source of error was negligible (corrections <1%). For thicker targets, heavy target elements and low γ-energies corrections had to be partially applied. Since we evaluated our spectra for energies down to 60 keV we had to apply corrections for this effect. This was done using the tables for photon attenuation coefficient taken from [cf. **111**]. Based on these data we calculated in dependence of the mass of the foil and the energy-dependent absorption coefficient (μ) for each single peak in a spectrum the fraction $r(E_\gamma)$ of γ-quanta, which leave the foil and can be registered by the detector by Eq. (3-12), by dividing each net-peak area by $r(E_\gamma)$ we took into account this effect.

$$r(E_\gamma) = \frac{1}{d}\int_0^d \exp(-\mu\cdot x)\,dx = \frac{1-\exp(-\mu\cdot d)}{\mu\cdot d} \qquad (3-12)$$

For all measurements of low energy lines, special low attenuation sample holders were used and correction for self-absorption in the metal foils was applied. We could rewritten Eq. (3-13) in the form as:

$$A(E) = A_0(E)\cdot\frac{\rho\cdot\left(\mu(E)/\rho\right)\cdot d}{1-\exp\left[-\rho\cdot\left(\mu(E)/\rho\right)\cdot d\right]}, \qquad (3-13)$$

Where, $A_0(E)$ is the measured activity for low energy lines at energy E, A(E) is the corrected activity, d is the thickness of the foil (μm), ρ is the density of the foil (g/cm^3), and (μ(E)/ρ) is the mass attenuation coefficient for the photon energy E, taken from. [cf. **111**]

3-8.6 Uncertainties due to Interfering Processes

Nuclide production as a consequence of severe interference by secondary neutrons produced in the stacks was observed in experiments caused partially large corrections. The cross sections derived from targets irradiated accounting for this contribution agreed within experimental uncertainties with those from single thin measurements. A method was developed to describe quantitatively the influence of secondary particles produced in the target stacks and to correct the measured data for their interference [cf. **22,110**]. The additional uncertainty caused by the correction for each individual reaction is considered in the errors given for the cross sections.

Due to the high cross section for all individual isotopes included in natural element of Mo, and Gd. Especial attention paid to the reactions between incident particles and target nuclei. Secondary particles are produced in the target could produce residual nuclei, this phenomenon is obviously clear above 200 MeV, imply a special care had to be taken to avoid its contributions. Secondary neutrons could formerly produce a significant interference Michel et al., (1982) [cf. **22**]; Schiekel et al., (1996) [cf. **112**]. Still at low and medium energy this attention must be paid for that reaction accompanied with high cross section for integrated secondary particle "neutrons" in our case which in course have a high interacting cross section at low energy, especially in presence of moderating agents. Nuclide production by medium-energy protons irradiating thick or extended targets is extremely complicated [cf. **109**]. By reactions of the primary particles a variety of secondary particles is produced which themselves are responsible for most of the production of residual nuclides. This production of secondary neutrons, protons, and light complex particles depends on the energy of the primary particles and on the bulk chemical composition of the irradiated target.

In order to describe nuclide production at a depth d inside a thick or extended target with a vector of bulk chemical constituents $C_b = (C_{b,L}, L=1,\ldots,n)$ and a size parameter R irradiated with protons having an initial energy $E_{p,ini}$. We have to calculate the transport of primary particles inside the target as well as the production and transport of secondary particles. In order to derive the differential flux densities $dJ_k(E_k, d, R, C_b, E_{p,ini})/dE_k$ of all primary and secondary particles k which themselves can undergo nuclear reactions. Then the production rate of a residual nucleus i in a special part of the target having $C_s=(C_{s,j}, J=1,\ldots,n)$ can be calculated by M. Gloris et al., (2001) [cf. **109**] as follows:

$$P_i\left(d,R,C_s,C_b,E_{p,ini}\right)=N_A\sum_j\frac{C_{s,j}}{A_j}\sum_k x\int_0^{E_{p,ini}} dE_k\sigma_{i,j,k}\left(E_k\right)\frac{\partial J_k}{dE_k}\left(E_k,d,R,1,E_{p,ini}\right) \qquad (3-14)$$

In Eq. (3-14), the $\sigma_{i,j,k}(E_k)$ are the thin target cross sections for the production of a residual nucleus i by particle type k from the target element j. N_A is Avogadro's number, and A_j the atomic mass of target element j.

The difference between thin, thick, and extended targets can be coarsely defined on the basis of the classification of targets according to their size parameter τ in terms of the interaction length μ [cf. **109**].

$$\mu = \sum_{l=1}^{n} \left(\frac{A_l}{N_A \cdot \sigma_{a,l}} \right) \qquad (3-15)$$

We speak-of thin targets if $\mu >> \tau$, of thick targets if $10\mu > \tau > 0.1\ \mu$, and of extended targets if $\tau >> \mu$. For protons, typical interaction length are $\mu(O) = 84\ g \cdot cm^{-2}$, $\mu(Fe) = 127\ g \cdot cm^{-2}$, $\mu(Bi) = 196\ g \cdot cm^{-2}$, and $\mu(U) = 207\ g \cdot cm^{-2}$ for oxygen, iron, bismuth, and uranium target, respectively. If one uses the geometric cross-section as a crude approximation of the absorption cross section $\sigma_{a,\,l}$.

In thin targets the contributions of secondary particles to nuclide production can mostly be neglected. In thick or extended targets, primary and secondary neutrons cause most of the production of residual nuclides. Especially in cases sensitive producing of secondary particle and interacting of them with bulk targets. In the experiments the following approach was used to estimate the influence of secondary neutrons. According to Gilbert et al., (1993) [cf. **110**] a differential flux density $d\Phi_n/dE_n$ was estimated by introducing a cobalt foil in the middle and the end of the stacks. In these foils ^{59}Fe was generated and measured which is produced exclusively by secondary neutrons via $^{59}Co(n,p)^{59}Fe$ with a threshold energy of 0.8 MeV (Q-value –0.78 MeV):

$$\frac{d\Phi_n}{dE_n} = \Phi_0 \qquad for\ E_n \in [1, 3.13\ MeV]$$

$$\frac{d\Phi_n}{dE_n} = \Phi_0 / E_n \qquad for\ E_n \in [3.13, E_{p,prim}\ MeV] \qquad (3-16)$$

where Φ_0 is a normalization constant according to Gilbert et al., (1993) [cf. **110**]. The contribution of secondary neutrons was estimated with determined Φ_0 via:

$$F = \int_1^{E_{p,prim}} \frac{d\Phi_n}{dE_n} \cdot \sigma_n(E_n)\, dE_n \qquad (3-17)$$

With F being the response integral by secondary neutrons and $\sigma_n(E_n)$ being the cross section for the neutron induced reaction. As $\sigma_n(E_n)$ experimental data from EXFOR database were used as far as available. Otherwise they were calculated with the ALICE-91 or EMPIRE-II codes Blann et al., (1988); Herman et al., (2005) [cf. **29**,**30**]. Values for Φ_0 were derived from the measured ^{59}Fe in the Co-foils. Then the expected neutron induced responses were exemplary calculated for a reaction with a low threshold, i.e. $^{89}Y(n,2n)^{88}Y$.

3-8.7 Uncertainties due to Random Coincidence Summing

The non-linear dependence of a photo-peak count rate measured with a multichannel analyzer on the actual source strength of the considered nuclide arises from summing effects.

True summing of γ-rays emitted in cascade is rate-independent and need not be considered in the case of comparative activation analysis (where the same nuclide is measured in different samples). Opposed to this is random summing (often-termed pulse pile-up). Where the pulse of interest sums with a pulse originating from a genetically not related γ-ray and as a consequence is lost from the photo-peak. This effect is rate-dependent and must always be taken into account except for the very seldom realized situation where sample and standard source have the same total source strength. Random summing takes place whether the source considered is composed of only a single nuclide or a mixture of several nuclides. Especially, errors due to random summing where there is usually a large difference between the total source strengths of the sample and standard used in detector efficiency calibration.

At proton energies above 10 MeV, due to the thresholds that opening of several other strong reaction channels among difference isotopes of natural abundant targets, a further correction may become mandatory. For a single reaction which, dominates in presence of other reactions cross section is low and since a radiochemical separation of radionuclides concerned from sample matrix activity could not be applied, it was necessary to measure the irradiated sample close to the detector. The high activity of samples probably sometimes caused a significant dead time, which could not be avoided decreased with time. A correction for the changing dead time was therefore developed, represented in the form:

$$D(t) = D(0) \cdot \exp(-\lambda^* \cdot t) \qquad (3-18)$$

where D(t) is a correction factor at time t, D(0) is the value of the factor at the end of irradiation, and λ^* is the decay constant which causes the dead time.

By using a pulse generator of constant frequency connected to the preamplifier we could determine the averaged dead times of the measurements or with previously determined system time constant, for approximation equal double value of amplifier time constant A. Wyttenbach, (1971); F. Cserpak, et al., (1994) [cf. **107,108**]. During the course of repeated spectrum measurements after an irradiation (mentioned above), dead times were also registered. Fitting a curve to those measured dead-time values we got D(0) and λ^* values. In general, for measured dead times of less than 10%.

Random summing is due to the finite time resolution of the detector system. The condition for a pulse not to be summed is that. This pulse is preceded and followed by a certain time τ into which no other pulse falls. Since, the time distribution of pulses originating from the detection of γ-rays emitted by a radioactive source is governed by a Poisson distribution, the probability p_1 of having no pulse in the time interval τ is:

$$p_1 = \exp(-N \cdot \tau) \qquad (3-19)$$

where, N is the mean rate of pulses emitted by the detector. It follows that the probability p_2 for not observing a random coincidence is

$$p_2 = p_1 \cdot p_1 = \exp(-2N \cdot \tau) \qquad (3-20)$$

where, the probability p_3 for observing a random coincidence.

$$p_3 = 1 - p_2 = 1 - \exp(-2N \cdot \tau) \qquad (3-21)$$

Since a random coincidence entails the loss of a pulse from the photo-peak, p_3 gives the fractional loss suffered by this photo-peak:

$$\frac{I_0 - I}{I_0} = p_3 = 1 - \exp(-2N \cdot \tau)$$

And

$$\frac{I}{I_0} = \exp(-2N \cdot \tau) \qquad (3-22)$$

Where,

I_0 photopeak count rate of the nuclide i, measured from a source of negligible strength,

I photopeak count rate of the same amount of nuclide i, measured as a component of a source of the total pulse rate N.

Supposing that the time interval τ is not a function of the pulse amplitude, it follows from Eq. (6) that all photo-peaks of a given spectrum suffer the same fractional loss. It can also readily be seen that a given photopeak in two different spectra suffers the same loss only when the total source strengths of these two spectra are equal. Since this condition is hardly ever realized in practical work, one is forced to use the count rate I_0 when comparing the intensities of a given photopeak in several different spectra.

One way to obtain I_0 is to feed a pulser with known repetition rate C_0 into the preamplifier and to measure its peak count rate C in the spectrum. When the pulses of the pulser are randomly distributed with respect to the pulses of the source, a relation identical to Eq. (3-22) holds:

$$\frac{C}{C_0} = \exp(-2N \cdot \tau) \qquad (3-23)$$

And from Eqs., (3-22) and (3-23) we have

$$C_0 = C \cdot \frac{I_0}{I} \qquad (3-24)$$

A procedure working along these lines has shown that there is a negligibly small correction term to Eq.(3-37), which is due to the fact that pulses cannot sum with each other. Another way to obtain I_0 is to estimate N, which can be done simply by integrating over the whole spectrum, and to apply Eq. (3-35) directly. The resolving time τ, the knowledge of which is required, can be measured with a double pulser; alternatively, it can be evaluated by measuring the photopeak count rate as a function of N. Since

$$\frac{I}{I_0} \approx (-2N \cdot \tau) \qquad (3-25)$$

Equation (3-25) is the first order approximation to Eq. (3-23), a plot of $(I_0-I)/I_0$ vs. N gives the value of τ. In the detector system used in this work, τ turned out to be somewhat larger than double the shaping time of the amplifier (2.5 μsec against 1 μsec. Instead of determining N by integration, we found it more practical to determine the duration of the measuring interval in both the live-time and true-time modes; when using a multichannel analyzer, these time intervals are related by:

$$T - t = \left[t \cdot (\alpha + \beta) \right] \qquad (3-26)$$

where
T- measuring interval (true time); t- measuring interval (live time)
α- fixed conversion time of memory cycle,
β- mean channel-dependent conversion lime of the memory cycle.
For a given spectrum, β will be constant, and Eq. (3-26) reduces to

$$T - t = \left[t \cdot (N \cdot \theta) \right] \qquad (3-27)$$

With

$$\theta = (\alpha + \beta) \qquad (3-28)$$

Combining Eqs (3-25) and (3-27) we get:

$$\frac{I}{I_0} = 1 - \frac{2 \cdot \tau}{\theta} \cdot \left(\frac{T}{t} - 1 \right) \qquad (3-29)$$

and finally

$$I_0 = I \cdot \left[1 - \frac{2 \cdot \tau}{\theta} \cdot \left(\frac{T}{t} - 1 \right) \right]^{-1} \qquad (3-30)$$

As can be seen from Eq. (3-30), the correction for random coincidence losses can be made in a first order approximation by measuring the counting interval simultaneously in the true-lime and live-lime modes. The factor 2τ/θ, the knowledge of which is required to make the correction, can be obtained from a linear plot of I/I_0 (or Y/Y_0) vs. T/t.

Coincidence losses may introduce a considerable error in γ-ray spectrometry and are thus a serious potential source of inaccurate activation analysis. A simple first order correction for coincidence losses can be applied by measuring the counting interval in both the true-time and in the live-time modes; the proposed equation is Shown by two experiments to hold up to an integral count rate of at least 4000 cps. The influence of high count rates on the resolution of HPGe or Ge(Li) detectors is a well known phenomenon. Contrary to this, the influence of high count rates on the measured intensity of a γ-line is little appreciated; there is hardly any publication on proton induced activation reactions using gamma spectrometers (where high count rates are often encountered) that pays attention to the difficulties of measuring correct intensities. This situation is very unsatisfactory, since to our experience the error in the intensity determination is in many cases much greater than the other errors involved in the spectra analysis. We want to pay an attention to this situation and propose a simple first order correction for coincidence losses.

3-8.8 Uncertainties due to True Coincidence

True-coincidence occurs in measurements involving a radionuclide that emits multiple cascade radiation. When two or more cascading photons give rise to a total or partial energy deposition in the detector within the response (or resolving) time of the detecting system, coincident detection of these photons appears. True-coincidence may result in summing effects or loss effects in the forms of γ–γ, γ–KX (IC), γ–KX (EC), etc. [cf. **90**]. The significance of these effect increases with increasing counting efficiency, therefore, the correction for true coincidence effects is of vital importance. We will deal with this matter first in a semi-empirical approach just for illustration in primary simple stage, then will turn to using a Monte Carlo treatment.

γ–γ Coincidences losses or gains in the net-peaks of γ-lines due to γ–γ-coincidences have to be taken into account at very small distances between sample and detector, and for some nuclides decaying by β^+-decay the 511 keV γ-radiation caused some problems. Since distances from the detector down to 5 cm were used to get a sufficient counting statistics, it would have been necessary to correct them in some cases depending on the measuring geometry. However, the corrections were in general less than 1% if one carefully avoids γ-transitions with high coincidence probabilities.

The true-coincidence effects for extended volume sources is more complicated than the situation of point sources, since the interactions of radiation within the source itself has to be taken into consideration. In this case peak and total efficiencies have to be well known not only as a function of photon energy but also as a function of space vector inside the source volume (V) According to [cf. **105**,**106**]. In order to circumvent the tedious integration over $\varepsilon_p(r)$, and $\varepsilon_t(r)$ at each differential element dV at r, a volume effect factor F_v needs to be introduced.

Let us first consider a radionuclide with a simple decay scheme as shown in Figure (3-18). The probability for γ–γ coincidence summing A=B+C can be expressed as:

$$S(\underline{A}=B+C)=\frac{\gamma_B}{\gamma_A}\cdot a_c\cdot c_c\cdot\frac{\int_V \varepsilon_{p\cdot B}(r)\cdot\varepsilon_{p\cdot C}(r)\cdot dV}{\int_V \varepsilon_{p\cdot A}(r)\cdot dV} \qquad (3-31)$$

$$S(\underline{A}=B+C)\equiv\frac{\gamma_B}{\gamma_A}\cdot a_c\cdot c_c\cdot\frac{\overline{\varepsilon}_{p\cdot B}\cdot\overline{\varepsilon}_{p\cdot C}}{\overline{\varepsilon}_{p\cdot A}}\cdot F_V(B,C) \qquad (3-32)$$

Where, γ is the absolute gamma intensity, a is the branching ratio; $c=1/(1+\alpha_1)$, $\alpha_1=\alpha_K+\alpha_L+....$, is the total internal conversion coefficient.

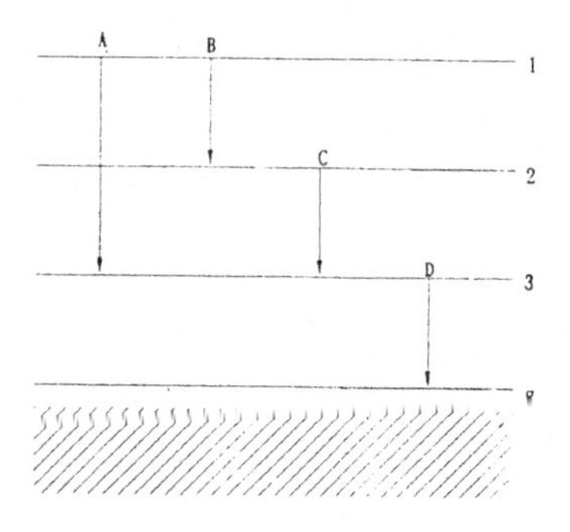

Figure (3-18): A simple Decay Scheme Showing (a) True-Coincidence Summing A=B+C. (b) True-Coincidence Loss γ–γ and γ–KX (IC), and (c) True-Coincidence Loss γ–γ and γ–KX (EC) Effects.

Consider the simple γ–γ and γ–KX (IC) coincidence as shown in Figure (3-18). Similarly, the probability for coincidence loss can be expressed as:

$$L(A)=L(A-B)+L(A-KX(B))$$
$$=a_B\cdot c_B\cdot\overline{\varepsilon}_{t,B}\cdot F_V(A,B)+a_B\cdot c_B\cdot\alpha_{K,B}\cdot\omega_K\cdot\sum_I k_i\cdot\overline{\varepsilon}_{t,B}\cdot F_V(A,k_i) \qquad (3-33)$$

With $C_B\alpha_{K,B}$ the probability that a K-electron is emitted, ω_K the fluorescence yield of the KX-ray, k_i, the relative emission rate and i can be α_1, α_2, etc.

The $\bar{\varepsilon}_p$ and $\bar{\varepsilon}_t$, in the above two equations are the source-volume-averaged peak and total efficiencies, which can be determined by efficiency calibrations. (In the following text ε_p and ε_t, will be used to replace the symbols $\bar{\varepsilon}_p$ and $\bar{\varepsilon}_t$, for simplicity).

The above reasoning based on simple decay schemes Figure (3-18) can be extended to more complicated decay schemes.

3-8.8.1 γ–γ Coincidence for Extended Sources

To illustrate the correction of true-coincidence effects, let us consider a radionuclide with a simple decay scheme as shown in Figure (3-18) where the gamma line of photon A is to be measured. The rate of pulses from full-energy absorption of photon A in the detecting system is:

$$N_{p,A} = I \cdot \gamma_A \cdot \varepsilon_{p,A} \qquad (3-34)$$

with I = source activity; γ = absolute gamma intensity; ε_p = full-energy peak efficiency.

Now, if we take into consideration the γ–γ coincidence summing $\bar{A}$ = B + C. Figure (3-18), the recorded full-energy peak in the spectrum would become:

$$N'_{p,A} = I \cdot \gamma_A \cdot \varepsilon_{p,A} + I \cdot \gamma_B \cdot \varepsilon_{p,B} \cdot a_C \cdot c_C \cdot \varepsilon_{p,C}, \qquad (3-35)$$

$$N'_{p,A} = I \cdot \gamma_A \cdot \varepsilon_{p,A} \cdot \left[1 + \frac{\gamma_B}{\gamma_A} \cdot a_C \cdot c_C \cdot \frac{\varepsilon_{p,B} \cdot \varepsilon_{p,C}}{\varepsilon_{p,B}}\right] = N_{p,A}\left[1 + S(\bar{A})\right] \qquad (3-36)$$

where S($\bar{A}$) is the probability for full-energy peak gains of photon A due to γ–γ coincidence summing $\bar{A}$ = B+C:

$$S(\bar{A}) = S(\bar{A} = B + C) = \frac{\gamma_B}{\gamma_A} \cdot a_c \cdot c_c \cdot \frac{\varepsilon_{p,B} \cdot \varepsilon_{p,C}}{\varepsilon_{p,A}} \qquad (3-37)$$

Where,

a = branching ratio

= the probability for the transition from one energy level to another (e.g., c_c gives the probability for the transition from level 2 to level 3 in Figure (3-18):

c = conversion factor

= the probability that gamma photon is emitted during the transition, i.e., internal conversion does not take part during the transition (e.g., c_c gives the probability that gamma photon B is emitted in Figure (3-18));

ε_p = full-energy peak efficiency.

If other series of cascading gamma lines (not shown in Figure (3-18) are also present, e.g., $\overline{A}$ = M+N+ O, one should also consider:

$$S\left(\overline{A}=M=B+C\right)=\frac{\gamma_M}{\gamma_A}\cdot a_N\cdot c_N\cdot a_O\cdot c_O\cdot\frac{\varepsilon_{p,N}\cdot\varepsilon_{p,N}\cdot\varepsilon_{p,O}}{\varepsilon_{p,A}}\qquad(3-38)$$

Leading to:

$$S\left(\overline{A}\right)=S\left(\overline{A}=B+C\right)+S\left(\overline{A}=M+N+O\right)$$
$$+S\left(\overline{A}=other\ possibl\ sum\ gammas.\right)\qquad(3-39)$$

If we consider the γ–γ coincidence loss $\overline{A}$ – D, as shown in Figure (3-18) the recorded full-energy peak in the spectrum would become:

$$N'_{p,A}=I\cdot\gamma_A\cdot\varepsilon_{p,A}-I\cdot\gamma_A\cdot\varepsilon_{p,A}\cdot a_D\cdot c_D\cdot\varepsilon_{t,D},\qquad(3-40)$$

$$N'_{p,A}=I\cdot\gamma_A\cdot\varepsilon_{p,A}\cdot\left[1-a_D\cdot c_D\cdot\varepsilon_{t,B}\right]=N_{p,A}\left[1-L\left(\overline{A}\right)\right]\qquad(3-41)$$

Where, L($\overline{A}$) is the probability for full-energy peak losses of photon A due to γ–γ coincidence loss $\overline{A}$ –D:

$$L\left(\overline{A}\right)=L\left(\overline{A}-D\right)=a_D\cdot c_D\cdot\varepsilon_{t,D}\qquad(3-42)$$

with ε_t, the total detecting efficiency. If other coincidence losses (not shown in Figure (3-18) are also present, e.g., $\overline{A}$–F. one should also consider:

$$L\left(\overline{A}-F\right)=a_F\cdot c_F\cdot\varepsilon_{t,F}\qquad(3-43)$$

Leading to

$$L\left(\overline{A}\right)=L\left(\overline{A}-D\right)+L\left(\overline{A}-F\right)+L\left(\overline{A}=other\ possible\ gammas\right)\qquad(3-44)$$

The above formulae can be extended to more complicated decay schemes, details have been reported elsewhere [cf. **105**,**106**].

For extended volume sources, the true-coincidence effects become more complicated, since the differential peak- and total-efficiency distributions within the source itself have to be taken into consideration. In this case simple formulae such as from Eq. (3-43) to Eq. (3-50) are no longer valid, and peak and total efficiencies have to be known not only as a function of photon energy but also as a function of space vector inside the source volume *V*.

Considering the γ–γ coincidence summing $\bar{A} = B - C$, Figure (3-18) for an extended source. Eqs. (3-41), and (3-42) have to be rewritten in their differential forms:

$$n_{p,A} \cdot dV = i(r)\gamma_A \cdot \varepsilon_{p,A}(r) \cdot dV \qquad (3-45)$$

$$n'_{p,A} \cdot dV = i(r)\gamma_A \cdot \varepsilon_{p,A}(r) \cdot dV + i(r)\gamma_B \cdot \varepsilon_{p,B}(r) a_C \cdot c_C \cdot \varepsilon_{p,C} \cdot dV \qquad (3-46)$$

where i(r) is the source activity density, $\varepsilon_p(r)$ and $\varepsilon_t(r)$ the peak and total efficiencies at *r*. If the radionuclides are homogeneously distributed in the source with volume V. *i(r)* = *I/V,* and by integration one may have:

$$N_{p,A} = \frac{1}{V} \cdot \gamma_A \cdot \int_V \varepsilon_{p,A} \cdot dV \qquad (3-47)$$

$$N'_{p,A} = N_{p,A} \left[1 + \frac{\gamma_B}{\gamma_A} \cdot a_C \cdot c_C \cdot \frac{\int_V \varepsilon_{p,B} \cdot \varepsilon_{p,C} \cdot dV}{\int_V \varepsilon_{p,A} \cdot dV} \right]. \qquad (3-48)$$

Therefore, the probability for γ–γ coincidence summing S is appearing in Eqs. (3-37), should be modified as:

$$S\left(\bar{A} = B + C\right) = \frac{\gamma_B}{\gamma_A} \cdot a_C \cdot c_C \cdot \frac{\int_V \varepsilon_{p,B} \cdot \varepsilon_{p,C} \cdot dV}{\int_V \varepsilon_{p,A} \cdot dV} \qquad (3-49)$$

Similarly, the probability for γ–γ coincidence loss *L* is appearing in Eqs. (3-33), should be modified as:

$$L(\overline{A}-D)=a_D\cdot c_D\cdot\frac{\int_V \varepsilon_{p,A}\cdot\varepsilon_{t,D}\cdot dV}{\int_V \varepsilon_{p,A}\cdot dV} \qquad (3-50)$$

Please note that the $\varepsilon_{p,A}(r)$ term now appears in both the numerator and the denominator in Eq. (3-56) and cannot be cancelled one with the other.

The summing and loss probabilities for more complicated coincidences can be obtained based on the same reasoning as above.

The determination of S and L values, in the forms such as Eqs. (3-49), and (3-50), needs information about $\varepsilon_{p,A}(r)$ and $\varepsilon_{t,A}(r)$ at each differential volume element *dV* at r, for each counting geometry and for each specific photon energy. It is very impractical and an almost impossible mission for most laboratories to obtain this information by experimental measurements. To circumvent tedious and immeasurable experimental efforts involved in *S* and *L* determination, a computational method is suggested in this work by introducing a "volume-effect factor" as will be discussed in the following subsection.

3-8.8.2 The Volume-Effect Factor F_V

In this work, we define $\overline{\varepsilon}_{p,A}(r)$ and $\overline{\varepsilon}_{t,A}(r)$, the source-volume-averaged peak and total efficiencies, $\xi_p(r)$ and $\xi_t(r)$ the normalized distribution functions of peak and total efficiencies, then for a given photon Y, we may have:

$$\varepsilon_{p,Y}(r)=\overline{\varepsilon}_{p,Y}\cdot\xi_{p,Y}(r) \qquad (3-51.1)$$

$$\varepsilon_{t,Y}(r)=\overline{\varepsilon}_{t,Y}\cdot\xi_{t,Y}(r) \qquad (3-51.2)$$

Where,

$$\overline{\varepsilon}_{p,Y}=\frac{1}{V}\cdot\int_V \varepsilon_{p,Y}(r)\cdot dV \qquad (3-52.1)$$

$$\overline{\varepsilon}_{t,Y}=\frac{1}{V}\cdot\int_V \varepsilon_{t,Y}(r)\cdot dV \qquad (3-52.2)$$

$$\frac{1}{V}\cdot\int_V \varepsilon_{p,Y}(r)\cdot dV\equiv 1, \qquad (3-53.1)$$

$$\frac{1}{V}\cdot\int_V \varepsilon_{t,Y}(r)\cdot dV\equiv 1, \qquad (3-53.2)$$

Now, Eq. (3-61) can be re-written as:

$$S\left(\overline{A}=B+C\right)=\frac{\gamma_B}{\gamma_A}\cdot a_C\cdot c_C\cdot\frac{\overline{\varepsilon_{p,B}}\cdot\overline{\varepsilon_{p,C}}}{\overline{\varepsilon_{p,A}}}\cdot\frac{1}{V}\int_V\xi_{p,B}(r)\xi_{p,C}(r)dV$$

$$=\frac{\gamma_B}{\gamma_A}\cdot a_C\cdot c_C\cdot\frac{\overline{\varepsilon_{p,B}}\cdot\overline{\varepsilon_{p,C}}}{\overline{\varepsilon_{p,A}}}\cdot F_V \qquad (3-54)$$

Where, F_V is the volume-effect factor and is defined as:

$$F_V\left(B_p,C_p\right)\equiv\frac{1}{V}\int_V\xi_{p,B}(r)\xi_{p,C}(r)dV. \qquad (3-55)$$

Similarly, Eq. (3-56) can be re-written as:

$$L\left(\overline{A}-D\right)=a_D c_D\overline{\varepsilon_{t,D}}\cdot\frac{1}{V}\int_V\xi_{p,A}(r)\xi_{t,D}(r)dV,$$

$$=a_D c_D\overline{\varepsilon_{t,D}}\cdot F_V. \qquad (3-56)$$

With,

$$F_V\left(A_p,D_t\right)\equiv\frac{1}{V}\int_V\xi_{p,A}(r)\xi_{t,D}(r)dV \qquad (3-57)$$

Please note that if the volume-effect F_V factor is removed then Eqs. (3-54), and (3-55) reduce to Eqs. (3-49), and (3-50). In other words, F_V is a factor that accounts for the fact that true-coincidence effects are actually depending on not only the source-volume-averaged peak and total efficiencies but also the differential-efficiency distributions within the source, if extended sources are encountered.

Where ε_p and ε_t, have a similar variation with distance to detector, and that the propagation of the uncertainty in the ε_t value to the COI value is of minor importance, in this work. We further assume:

$$\xi_{p,Y}(r)=\xi_{t,Y}(r)\equiv\xi_Y(r). \qquad (3-58)$$

With this assumption we have:

$$F_V\left(A_p,D_t\right)=F_V\left(A_p,D_p\right)\equiv F_V(A,D), \qquad (3-59.1)$$

$$F_V\left(B_p,C_p\right)\equiv F_V(B,C). \qquad (3-59.2)$$

This means F_V is only a function of $\xi_Y(r)$ combinations no matter ε_p or ε_t, is involved.

For coincidences involving more than two gammas energies, say Y_1, Y_2, Y_3, etc., we have:

$$F_V(Y1,Y2) = \frac{1}{V}\int_V \xi_{Y1}(r)\xi_{Y2}(r)dV, \qquad (3-60.1)$$

$$F_V(Y1,Y2,Y3) = \frac{1}{V}\int_V \xi_{Y1}(r)\xi_{Y2}(r)\xi_{Y3}(r)dV, \qquad (3-60.2)$$

$$F_V(Y1,Y2,Y3,Y4) = \frac{1}{V}\int_V \xi_{Y1}(r)\xi_{Y2}(r)\xi_{Y3}(r)\xi_{Y4}(r)dV, \qquad (3-60.3)$$

etc.,,

In summary, the probabilities for true-coincidence summing (S) and loss (L) can now be determined by (1) finding the γ. V, and c values from tabulated or reported nuclear data. (2) determining the ε_p and ε_t, values by efficiency calibration, and (3) determining the F_V Factors by calculation.

The ε_t-E_γ curves can be constructed by the conversion formula:

$$\varepsilon_t = \varepsilon_p \left(T/_P \right). \qquad (3-61)$$

Where P/T is the peak-to-total ratio, i.e. the ratio between full-energy peak area and the total area in the spectrum originating from the same decay transition. The P/T ratio is an experimentally determined quantity, and is a function of gamma-ray energy and the counting geometry.

The calculations of F_V factors were performed with the aid of the Monte Carlo program [cf. **100**]. This program can calculate the effective solid angle F_V for a specific energy of photon, for a specific source and counting geometry. The effective solid angle is:

$$\overline{\Omega} = \int_{detector}^{source} F_{att}\, F_{eff}\, d\Omega. \qquad (3-62)$$

Where, F_{eff} is the probability for a photon emitted within a differential geometrical solid angle $d\Omega$ and impinging on the detector to interact incoherently with the active detector crystal. F_{att} is the attenuation factor of the photon by the source itself, and by any material between the source and the detector crystal. Since $\overline{\Omega}$ is proportional to the peak efficiency ε_p [cf. **113**,**114**]. We can directly relate the distribution of $\overline{\Omega}(r)$ to the distribution function $\xi(r)$. From Eqs., (3-54) and (3-56) we have:

$$\left(\frac{1}{V}\right)\int_V \xi(r)\,dV = 1, \qquad (3-63)$$

The equation can be re-written as:

$$\left(\frac{1}{V}\right)\int_V C\cdot\overline{\Omega}(r)\,dV = 1, \qquad (3-64.1)$$

Where C is the normalization constant and can be expressed as:

$$C = \frac{V}{\int_V \overline{\Omega}(r)\,dV} = \frac{V}{E}, \qquad (3-64.2)$$

With

$$E = \int_V \overline{\Omega}(r)\,dV, \qquad (3-64.3)$$

And

$$\xi(r) = C\cdot\overline{\Omega}(r) = \frac{V}{E}\cdot\overline{\Omega}(r), \qquad (3-64.4)$$

Taking the F_V factor Y1-Y2 coincidence as an example and applying the above formulae, we have:

$$F_V(Y1,Y2) = \frac{1}{V}\int_V \xi_{Y1}(r)\,\xi_{Y2}(r)\,dV, \qquad (3-65.1)$$

$$= \frac{V}{E_{Y1}\,E_{Y2}}\int_V \overline{\Omega}_{Y1}(r)\,\overline{\Omega}_{Y2}(r)\,dV, \qquad (3-65.2)$$

Now, based on determined E and $\overline{\Omega}(r)$ values, F_V can easily be calculated using the above equation. The F_V factors for other coincidences can be calculated similarly.

3-8.8.3 Peak Area Correction

With the probabilities for γ–γ coincidence summing [S(A)] and loss [L(A)] determined, the observed full-energy peak area for photon A from a measurement $N'_{p,A}$ can be expressed as:

$$N'_{p,A} = N_{p,A}\left[1-L(\bar{A})\right]\cdot\left[1+S(\bar{A})\right]$$

$$= N_{p,A}\left[1-L(\bar{A})+S(\bar{A})-L(\bar{A})S(\bar{A})\right] \quad (3-66.1)$$

Or

$$N_{p,A} = {N'_{p,A}}\big/{COI}, \quad (3-66.2)$$

where $N_{p,A}$ is the corrected peak area and the coincidence correction factor COI is defined as:

$$COI = {N'_{p,A}}\big/{N_{p,A}} = \left[1-L(\bar{A})\right]\cdot\left[1+S(\bar{A})\right] \quad (3-66.3)$$

Could be re-written as:

$$COI^{\exp} = {\varepsilon_{p,t}^{COI}(E_\gamma)}\big/{\varepsilon_{p,p}^{free}(E_\gamma)}. \quad (3-66.4).$$

Where, $\varepsilon_{p,i}^{free}(E_\gamma)$ is the peak efficiency measured by using coincidence free sources, $\varepsilon_{p,i}^{COI}(E_\gamma)$ is the peak efficiency measured for gamma-lines E_γ from coincidence gamma-ray nuclide i.

It is worth mentioning here that the last term shown on the right hand side of Eq. (3-72.1), $L(\bar{A})S(\bar{A})$. Using the simple decay scheme of Figure (3-18) as an example, accounts for the fact that the sum pulses originating from B+C are subject to the same true coincidence loss (with D) as A itself [cf.**115**]. For other complicated decay schemes, the situations are the same. Therefore, we used the program "TRUECOINC" [cf. **116**] as a global form for number of cascades in the decay scheme data relieved from the isotope concerned data file.

True coincidence losses could be caused by every gamma ray in a cascade chain for the investigated one (g_0), which is emitted within the time resolution of the detector system. The k_{tcl} correction factor is the probability of the detection of a given g_0 gamma-quantum without the detection of any other gamma-quanta from its cascade chain. Further problem comes when during the de-excitation X-rays are emitted. If the detector is able to detect the low energy gamma, the X-rays also could cause true coincidence losses. X-rays are produced during the EC and the gamma conversion to electron processes. If the angular correlation between the cascading gamma is neglected this correction factor can be calculated as declared from the following equation by using "TRUECOINC". [cf. **116**]

$$\kappa_{tcl} = \frac{\sum_{i=1}^{n_0} \beta_i S_i \prod_{j=1}^{n_i} \left[(1-\varepsilon_{ij})\Gamma_{ij} + \sum_k \Gamma_{ijk}(1-\omega_k + \sum_l \delta_{kl}(1-\varepsilon_l)) \right]}{\sum_{i=1}^{n_0} \beta_i \prod_{j=1}^{n_i} \left[\Gamma_{ij} + \sum_k \Gamma_{ijk}(1-\omega_k + \sum_l \delta_{kl}) \right]} \qquad (3-67),$$

Where

- n_0 is the number of cascade chains containing the g_0 transition,
- n_i is the number of the cascade gamma transitions in the i-th cascade chain ($i=1,2,...,n_0$)
- b_i is the branching ratio of the β^- (or β^+, EC, α) transition populating the i-th cascade chain,
- G_{ij} is the branching ratio of the j-th gamma transition in the i-th cascade chain,
- G_{ijk} is the conversion ratio of the j-th transition trough creating vacancy on the k-th atomic shell (K,L1, L2…) in the i-th cascade chain,
- w_k is the fluorescence yield of the k-th atomic shell (K,L1, L2, …),
- d_{kl} is the emission ratio of l-th x –ray from the k-th atomic shell (K,L1, L2…),
- e_l is the product of the total detection efficiency and the self absorption for the energy of the l-th x-ray l= $(\varepsilon_t * \varepsilon_{sa})_l$,
- e_{ij} is the product of the total detection efficiency and the self absorption for the energy of the j-th gamma transition in the i-th cascade chain $\varepsilon_{ij} = (\varepsilon_t * \varepsilon_{sa})_{ij}$, (use $e_{ij}=0$ for the g_0 transition)
- S_i $S_i=1$ for β^- or a decay; and $S_i=[1-2*(\varepsilon_t * \varepsilon_{sa})_{511keV}]$ for β^+ decay while for EC $_i$.

$$\beta_i S_i = \sum_k \Delta_{ik}(1-\omega_k + \sum_l \delta_{kl}(1-\varepsilon_l)) \qquad (3-68)$$

- D_{ik} is the branching ratio of the EC transition trough generating vacancy on the k-th atomic shell (K, L1, L2…), while populating the i-th cascade chain.

True coincidence gains could also be occured if there are more levels between the levels of the investigated g_0, then the coinciding gamma rays can produce the same amplitude as the original transition.

$$\kappa_{tcg} = [1 + \frac{\sum_{i=1}^{n}\prod_{j=1}^{m_i}\Gamma_{ij}\varepsilon_p(\gamma_{ij})}{\Gamma_0\varepsilon_p(\gamma_0)}] \qquad (3-69)$$

In most cases this true Coincidence gain correction is negligible while the peak efficiency of the detectors is much less than the total efficiencies. In some case when G_0 is small the correction can be important.

Alias lines (or pseudo-lines) are non-existing gamma lines shown in the spectrum produced by the true coincidence of two or more non-neighboring gamma transition and X-rays. In the case of a complicated spectrum these alias line may coincide in energy with a real gamma line. Therefore the knowledge of the energies and relative intensity of these line is very important in the identification of the isotopes and to avoid the error in the data evaluation. The intensities are calculated by the following expression:

$$I_{alias,a,b} = \sum_{i=1}^{n_0}\beta_i\prod_{j=1}^{n_i}\left[\Gamma_{ij} + \sum_k \Gamma_{ijk}(1-\omega_k + \sum_l \delta_{kl})\right]R_a R_b \frac{\varepsilon_p(E_a)\varepsilon_p(E_b)}{\varepsilon_p(E_a+E_b)} \qquad (3-70)$$

where R_x is the branching ratio of the x type of transition to the total transition between the selected level. The calculation takes account only two coinciding lines.

3-8.9 Uncertainties of Fluence Response

Uncertainty of the flux densities were determined via the monitor reaction $^{nat}Cu(p,x)^{63}Zn$, $^{nat}Cu(p,x)^{65}Zn$, $^{nat}Ti(p,x)^{48}V$ using cross-sections as described in detail by Tarkanyi et al., (2001) in [cf. **94**]. The uncertainty of flux density is mainly determined by the uncertainties of efficiency and of the mass of the monitor foil, the monitor cross-sections, which sum up to be about 5%. Uncertainty of the measurement of the monitoring radionuclides, the relative precision of this determination was checked by multiple measurements and no error was assigned to the monitor cross section.

In this study we will deal with an approach to simply determine the beam intensity, and increasing the accuracy of data gained. An established method for measuring the neutron, proton fluence-rate and beam energy involves placing a standard sample in the beam and detecting the γ-rays given off by the reference reactions that take place.

To estimate the accurate flux through sample foils does not fullfilled with assigning the sources of flux uncertainties, and its correction magnitudes as menstioned above. In addition to that, we may consider for possible effects of secondary neutron activation to be ruled out to a large extent as reported by Weinreich et al., (1980) [cf. **117**]. The effects of that processes could be take place inside a single foil as well as a multiple foils used in stacked foils technique.

This approach is built to audit the beam intensity, to utilize from the stacked foils techinque in modeling its change. The degraded fluence, whatever it was proton or neutron beam is effected with several reasons Scattering which causes also an electric stopping power for charged particle, beam get out fraction which is responsible for nuclear stopping power, Straggling effects due to essential property of generated beam from charged particles accelerators. (i.e. emittance angle, focusing, and defocusing properties, etc.,). The effects of spaced media through which the beam pass before reach to stacked foils in case of internal or external irradiation. Figure (3-19) shows the neutron production cross section from proton induced reaction for natural Molybdenum, and Gadolinium target foils as a function of the proton energy, E_p up to 20 MeV, calculations performed using ALICE-91 Code [cf. **29**]. The Differential Cross-Sections for neutron spectrum generated from the natural Molybdenum, Gadolinium targets bombarded with 20 MeV Protons are shown in Figure (3-20).

We will deal with this dilemma by dividing the problem into two parts, first part will be evaluated experimentally and the second part will be dealt theoretically. The effective fluence through stacked foils will be determined, taking into account the different causes for beam degradation. The stack will be cured by a precisely determined flux using multiple monitors, then fitting these experimental points to theoretical formulae. It is proposed that the treatment will not remedy allover corrections but still some errors unresolved may cause the values of cross sections to be suffering deviation from absolute values.

In spite of, in the actual case where the product may be effected with other building up reactions make a partial contribution of second beam of seven orders less than incident beam. Especially when using a natural thick sample, the affinity of this second beam differs from the incident beam in type and reactivity (photon and particle induce reactions). Where the cross section for neutron is high especially with slowing down, through the interactions upon successive foil layers of target nuclei.

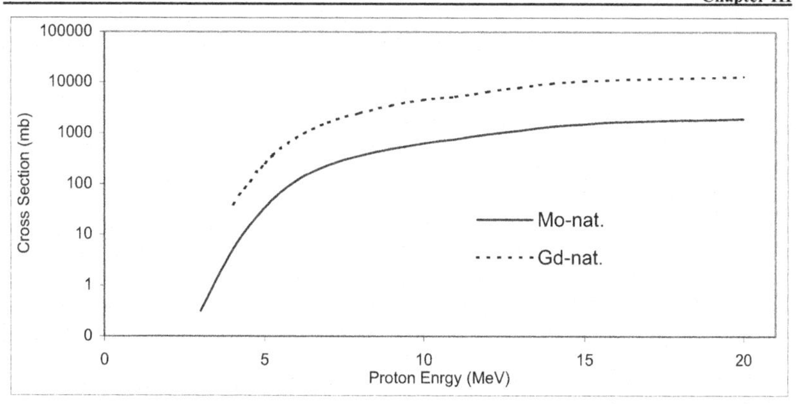

Figure (3-19): Neutron Production Cross Section from Proton Induced Reaction for Natural Gadolinium, and Molybdenum Target Foils as a Function of the Proton Energy, E_p up to 20 MeV, ALICE-91 Code [cf. **29**].

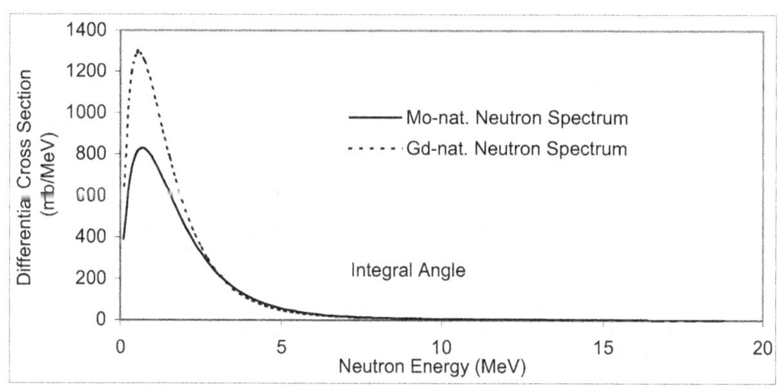

Figure (3-20): Differential Cross-Section Neutron Spectrum from Natural Gadolinium, and Molybdenum Target Foils Bombarded with 20 MeV Protons, ALICE-91 Code [cf. **29**].

The secondary beam build up will be remedied through calculating the produced nuclei during interaction of neutron along the stacked foils which can be approximated as an increase in number of target nuclei or as an increase in primary beam intensity. $F(E_i,t)$ represent the fluence response probability, probability means that percentage valued from 0 to 1 could be represence change in the specific fluence through one stacked foil. $F(E_i,t)$ can be expressed as a product of two functions (excluding primary beam buildup and subsequent neutrons). We write:

$$F(E_i,t) = f^{(1)}(E_i,t)\cdot f^{(2)}(E_i,t), \qquad (3-71)$$

3-8.9.1 $f^{(1)}(E_i,t)$, Neutron-Production Probability

In principle $f^{(1)}(E_i,t)$ depends upon all (p,xnγ) cross sections, where x=1,2,...etc. For the present purposes we assume that proton scattering cross section predominates. The assumption is reasonable because of the neutron spectrum resulting from the proton-induced reaction is considerably low. Furthermore, for proton energy between 1 and 4 MeV, the proton scattering cross section is more than 95% of the total (p,xnγ) cross section.

The discussion of the neutron-production probability is conveniently presented in terms of the differential probability, $(d/dx)[f^{(1)}(E_i,x)]dx$. It is defined, as the probability that a proton of energy E_i propagate from a target foil of thickness dx located at a depth x against Coulomb interaction, will give rise to a nuclear neutron due to absorbed proton per unit mass density per unit flux density. Functionally we write S.S. Malik (1975) [cf. **118**]:

$$\frac{\partial}{\partial x}\left[f^{(1)}(E_i,t)\right]dx = \frac{\left\{1-\exp\left[-n_t\cdot\sigma_{n,p}(E_i)\cdot x\right]\right\}\cdot n_t\cdot\sigma_{n,p}(E_i)}{\left\{\dfrac{t}{\Delta\left[R_p^0(E_i)-(t-x)\right]}\right\}}\cdot dx, \qquad (3-72)$$

Where, n_t is the number of target foil atoms per cm^3, and the total neutron production cross section. The neutron-production probability of practical interest is one, which contains the provision that the ejected neutron will reach the stacked foil end. The path lengths associated with such neutrons will depend upon a number of factors. Of these the finite sizes of the target foil contribute to the geometrical variation of the path lengths. A procedure, which incorporates the above provision and simplifies the treatment, is to write:

$$\frac{\partial}{\partial x}\left[f^{(1)}(E_i,t)\right]dx \rightarrow \frac{\left\{\exp\left[-n_t\sigma_{p,s}(E_i)x\right]\right\}n_t\sigma_{n,p}(E_i)}{\left\{\dfrac{t}{\Delta\left[R_p(E_i)-(t-x)\right]}\right\}}\cdot dx, \qquad (3-73)$$

Where, $\sigma_{p,s}(E_i)$ is the total proton scattering cross section

$$\sigma_{n,p}\left(E_i\right)=\frac{1}{4\pi}\int_0^{2\pi} d\phi \int_0^{\pi/2}\left(\frac{d\sigma_{n,p}}{d\Omega'}\right)\cos\psi \sin\psi \, d\psi , \qquad (3-74)$$

Eq. (3-79) indicates the probability that an incident proton of energy E_i propagate at a depth x in the range dx will pass through a stacked foil. $R(E_i)$ is the straight-ahead range of the proton in the material of the target foil. The Δ function has the property:

$$\Delta\left[R_p\left(E_i\right)-\left(t-x\right)\right]=1 \quad if\ R_p\left(E_i\right)\geq\left(t-x\right), \qquad (3-75)$$

$$=0 \quad if\ R_p\left(E_i\right)\prec\left(t-x\right). \qquad (3-76)$$

Through a straightforward integration yields:

$$\left\{\frac{\Delta\left[R_p\left(E_i\right)-\left(t-x\right)\right]}{t}\right\}\rightarrow R_p^0\left(E_i\right)\Big/ t \, . \qquad (3-77)$$

In defining the Δ function through the Eqs. (3-75) and (3-76) it has been tacitly assumed that there is no straggling.

It is further assumed that the range $R(E_i)$ is parallel to the axis of the target foil. The actual range $R(E_i)$ is however, not parallel to the target foil axis. The direction of the produced neutron changes due to collisions with the nuclei in the target foil.

In Eq. (3-74) $d\sigma/d\Omega'$ is the differential cross section per unit solid angle for the amount of neutron energy scattered in the direction ψ [cf. **118**]. Use of $\sigma_{n,p}(E_i)$ enables us to preferentially discriminate against those neutrons which possess small kinetic energy less and the associated neutrons need long path lengths to reach the stacked foils end. In addition it yields an ejected proton beam with an initial direction that is parallel to the incident proton beam. Values of $\sigma_{n,p}(E_i)$ per target atom are numerically computed.

In order to integrate Eq. (3-73) it is assumed that "a parallel proton beam of small cross-sectional area is incident axially upon a cylindrical target foil". This procedure assigns the same x value to all the protons present at a given depth in the target foil. The integration is now straightforward and we have:

$$f^{(1)}(E_i,t)=\int_0^t \frac{\partial}{\partial x}\left[f^{(1)}(E_i,t)\right]dx$$

$$=\left[\frac{R_p(E_i)}{t}\right]\cdot\left[1-\exp\left(-n_t\,\sigma_{p,s}\,t\right)\right]\frac{\sigma_{n,p}(E_i)}{\sigma_{p,s}(E_i)}. \qquad (3-78)$$

The neutrons produced in the stack could also be respomsible for the production of more extra neutron, the neutron production cross section from neutron induced reaction for natural Molybdenum, and Gadolinium target foils as a function of the neutron energy, E_n up to 5 MeV. The concerned data could be estimated by using ALICE-91 Code as represented in Figure (3-22).

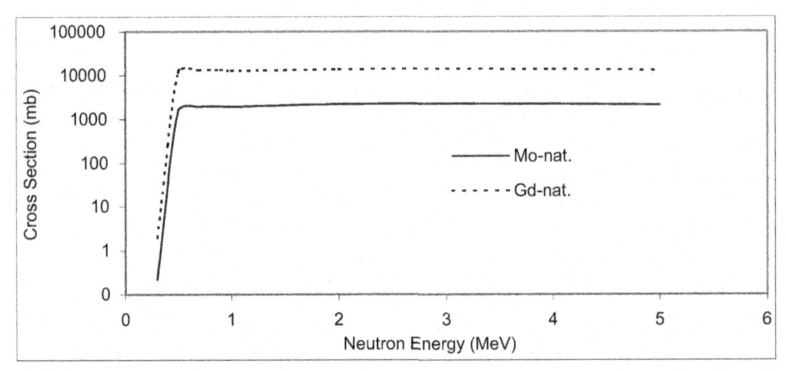

Figure (3-21): Neutron Production Cross Section from Neutron Induced Reaction for Natural Gadolinium, and Molybdenum Target Foils as a Function of the Neutron Energy, E_n up to 5 MeV, Using ALICE-91 Code.

The neutrons generated in the stack could produce radionuclide due to different types of reaction, depends on the reaction threshold and neutron energy. One of the most important reaction is the radiative capture cross section, which is generate gamma rays in the stack couls be responsible for driving more reactions produce neutrons by photoneutron reaction. Figure (3-22) represence some literatured experimental data concerning the capture cross section for natural target foils of Molybdenum, and Gadolinium as a function of neutron energy up to 5 MeV, which produces in the stack with high differential cross section as previously shown in Figure (3-20). In Figure (3-23) the relative values of the neutron production probability, $f^{(1)}(E_i,t)$, in arbitrary units, for Molybdenum, and Gadolinium target foil of thickness t= 25, and 10 μm, respectively as a function of the proton energy, E_p (MeV). We can see from the figure that the neutron production increases with the increase in the incident proton energy.

The production begins at threshold energy and increases with different values for Molybdenum and Gadolinium. Gd has the probability to produce neutrons more than Mo.

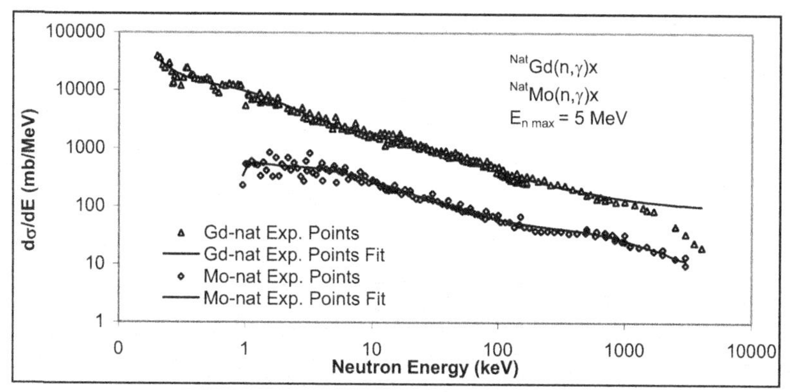

Figure (3-22): Neutron Induces Capture Cross Section for Natural Gadolinium, and Molybdenum Target Foils as a Function of the Neutron Energy, E_n up to 5 MeV in Log-Log Scale [cf. **119**, **120**].

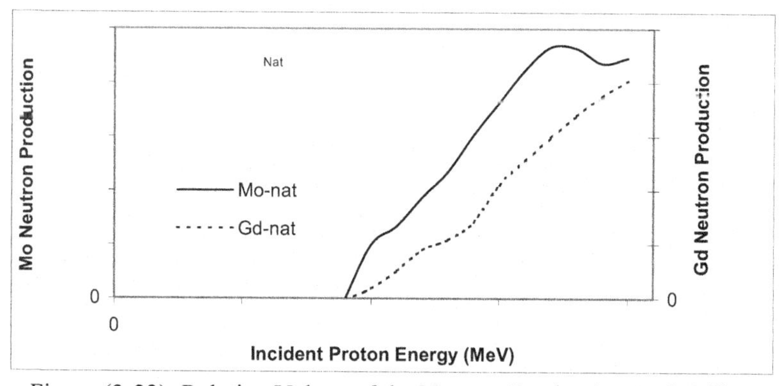

Figure (3-23): Relative Values of the Neutron Production Probability, $f^{(1)}(E_i,t)$, in Arbitrary Units, for Gadolinium, and Molybdenum Target Foils of Thickness t=10, and 25 μm, respectively as a Function of the Proton Energy, E_p (MeV).

3-8.9.2 $f^{(2)}(E_i,t)$, Neutron Transmission Probability

Once a produced neutron is emitted in the forward direction it must reaches the stacked foil end before it vanishes in a single target foil. We define the function $f^{(2)}(E_i,t)$, as the probability that a produced neutron of energy $E \leq E_i$ will pass through the target foil. The neutron transmission probability, providing that the neutrons will reach the stacked foils end will depend upon the total absorption cross section $\sigma_{n,a}(E_i)$. Due to the finite sizes of the target foil contribute to the geometrical variation of the neutron path lengths is minimum. A procedure, which incorporates the above provision and simplifies the treatment, is to write

We now introduce a differential probability for $f^{(2)}(E_i,t)$, and express it as:

$$\frac{\partial}{\partial x}\left[f^{(2)}\left(E_i,t\right)\right]dx \rightarrow \left\{\exp\left[-n_t\,\sigma_{n,s}\left(E_i\right)\right]x\right\}n_t\,\sigma_{n,a}\left(E_i\right)\cdot dx, \qquad (3-79)$$

Eq. (3-79) indicates the probability that a neutron produced at a depth x in the range dx (by a proton of energy E_i) will reach the stacked foil end. $\sigma_{n,s}(E_i)$ is the total scattering cross section for neutron, which affects the the neutron fluence in the material of the target foil.
A straightforward integration yields:

$$\begin{aligned} f^{(2)}\left(E_i,t\right) &= \int_0^t \frac{\partial}{\partial x}\left[f^{(2)}\left(E_i,t\right)\right]dx \\ &= \left[1-\exp\left(-n_t\,\sigma_{n,s}\,t\right)\right]\frac{\sigma_{n,a}\left(E_i\right)}{\sigma_{n,s}\left(E_i\right)}. \qquad (3-80) \end{aligned}$$

It has been tacitly assumed that there is no straggling. It is further assumed that the actual path for the neutrons is parallel to the normal axis of the target foil. The direction of the produced neutron changes due to collisions with the nuclei in the target foil. Subsequent to these collisions the scattered neutrons possess an angular distribution. If we assume that the number of collisions is sufficiently large (>20), the angular distribution is Gaussian about a certain mean direction. We take the direction of the scattered neutron to be that defined by the mean scattering angle, θ, on a plane parallel to the incident proton beam.

The value of θ for an neutron with initial kinetic energy E and for a foil of thickness t (surface density in gm/cm^2), atomic number Z and atomic weight A has been calculated by Paul C. Brand, NIST, (1999) [cf. **121**]. Using the values of θ are calculated as a function of neutron energy as:

$$\theta = \arcsin\left[\frac{4.52182}{t\sqrt{E(MeV)}}\right] \qquad (3-81)$$

The thickness x in the present calculations is not unique but varies between 0 and a length equal to the path $R(E_i)$ of the neutron in target foil. This necessitates averaging of the angle Ω in accordance with the formula:

$$\left[\Theta(E)\right]_{av} = \frac{1}{R}\int_0^R \Theta\, dx. \qquad (3-82)$$

The value of average path $R^0(E_i)$ is thus taken to be:

$$R_n^0(E_i) = R_n(E_i)\cos\left[\Theta(E)\right]_{av} \approx t, \qquad (3-83)$$

$$f^{(2)}(E_i,t) = \left[1-\exp\left(-n_t\cdot\sigma_{n,s}\cdot t\right)\right]\frac{\sigma_{n,a}(E_i)}{\sigma_{n,s}(E_i)}. \qquad (3-84)$$

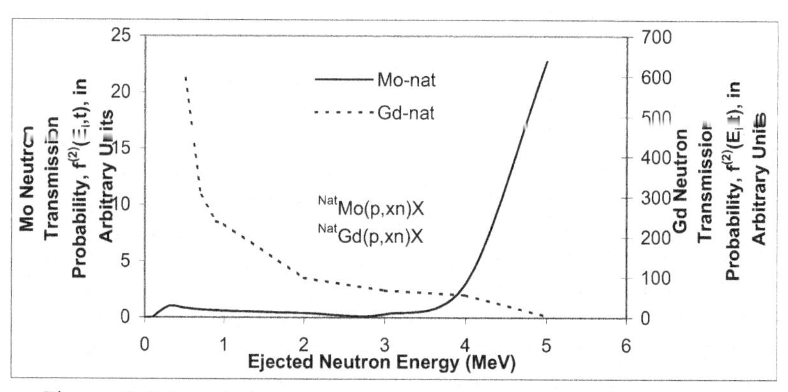

Figure (3-24): Relative Values of the Ejected Neutron Transmission Probability, $f^{(2)}(E_i,t)$, in Arbitrary Units, for Gadolinium, and Molybdenum Target Foils as a Function of the Ejected Neutron Energy, E_n (MeV).

The value $f^{(2)}(E_i,t)$ for ejected neutrons of energy E resulting from the interacting protons of energy E_i, is given in Figure (3-24) as a function of the incident proton energy. We can see the probability of neutron transmission for Molybdenum is higher than that for Gadolinium, this implies that Gadolinium could consume a large amount from the produced neutron, at the same time the Gd has the probability to produce neutron than Mo. This can be explained with that Gd could react with neutron through the radiative capture as shown in Figure (3-22), as well as (n,2n) reaction to produce more neutron with shifted energies toward the lower neutron energy. In this coarse, with respect to the secondary particles contribution in nuclear reaction producing readionuclides. We could use a combination of Mo, and Gd elements in certain arrangement for radionuclide production. In this arrangement the Molybdenum has the job to produce the neutrons and the Gadolinium has the job to moderate these neutron to a suitable energy needed.

We are now in a position to write down the overall expression for F(E$_i$, t),) the "Fluence Response Probability", Eq. (3-85) as:

$$F(E_i,t) = f^{(1)}(E_i,t)\, f^{(2)}(E_i,t) \qquad (3-71)$$

$$F(E_i,t) = \left[\frac{R_p^0(E_i)}{t}\right] \times$$

$$\cdot\left[\left[1-\exp\left(-n_c\,\sigma_{p,s}\,t\right)\right]\frac{\sigma_{n,p}(E_i)}{\sigma_{p,s}(E_i)}\cdot\left[1-\exp\left(-n_c\,\sigma_{n,s}\,t\right)\right]\frac{\sigma_{n,a}(E_i)}{\sigma_{n,s}(E_i)}\right]. \qquad (3-85)$$

This completes the discussion of the "Fluence Response Probability".

The quantity in the square brackets is plotted in figure (3-25) for Mo, and Gd. We can see the increase in both Mo and Gd as the incident energy increases. With respect to its behavior the fluence response probabilty for Gd is higher than Mo.Some qualifying remarks F(E$_i$,t) are appropriate at this juncture. For incident neutron with energy E>0.3 MeV the number of radioactive capture produced (specific primary capture) is only slightly dependent upon the energy of the neutrons. In our calculations it is not possible to assign any specific energy values to the neutron incident upon the successive foils.

Even if all the neutrons at their production had the same value of the kinetic energy, their energy distribution after passage through target foil will be fairly broad. This has the effect of dispersing the individual values that determining the overall value of activity. For incident neutrons having an energy E≤0.33 MeV the increase in the specific absorbing reaction is accompanied by the reduction of the effective thickness of the target foils. The number of absorbing captures produced by neutrons for medium to heavy nuclei with E<0.3 MeV would thus be increased on vice for light nuclei.

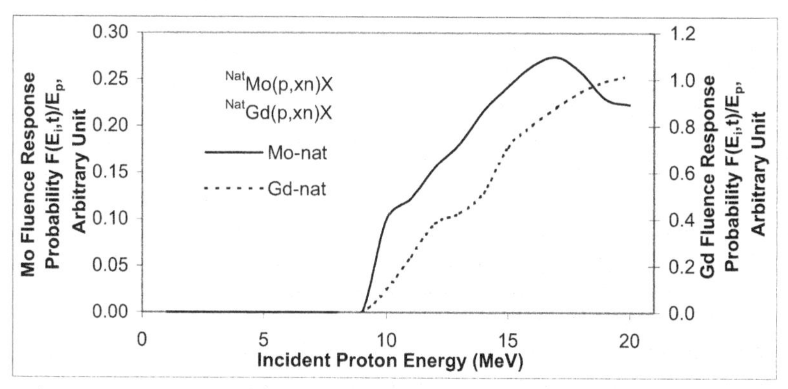

Figure (3-25): Relative Values of the Overall Proton Fluence Response Probability per unit Proton Energy, $F(E_i,t)/E_p$, in Arbitrary Units, for Gadolinium, and Molybdenum Target Foils, as a Function of the Incident Proton Energy, E_p (MeV).

The theory presented here is the first attempt, to our knowledge, to build a systematic theoretical structure of the sensitivity toward effective flux estimation, that is subsequent of radionuclide production of stacked foils. That implies considering the production cross section of monitor reaction, with the neutron production cross section of the sample. The neutrons produced itself represent a physical quantity for all-secondarys contribution, its spatial distribution, and transmission properties within the sample. We show that the combined effect of neutron production and neutron transmission through the target foil coupled with the primary, secondary beam, and may be the photonuclear activation also render the accuracy of the nuclear data measured with respect to the proton induced reactions. As we can see in molybdenum targets the neutron total production is high, for integral angles, and energies.

The cross section for neutron production approached two barn, with maximum at individual energy about 300 mb, at the same time the scattering cross section is high in the energy range of high population in neutron yield spectrum. From this we can imagine a yield affected with a consideration to those reasons, not only to the primary beam flux. At the same time gamma induces decay is a verified phenomenon, it appreciable to consider. If we looking for neutron as intermediate particle could be produced even with gamma induces decay of metastable state, or long lived radionuclides.

In brief as flux introduces as erroneous parameter imply a correction to remove the accompanied error, in turn the true one appear, this reflect to the number of count under the peak. As we know this count joint to activity through detector efficiency as a proportional constant. We nominate yttrium-89 for this mission because it has low threshold energy for neutron interaction, it can estimate the neutron content in the foil. The whole system then become has an unique efficiency that could varying the flux at different places is for sure effective and a number of probabilities represent all the corrections that could be considered during calculation of the reaction cross section, reaction yield.

3-9 Estimated Correction Values for Uncertainty Sources

From the corrected decay rates of the radioactive products and the measured beam currents, the cross sections of the proton induced nuclear reactions were calculated using the usual activation formula. The associated uncertainties are discussed below:

In the worst case all these uncertainties except for that of the net-peak counts summed up quadratically to about 10%. To this the latter individual uncertainty had to be added quadratically. As a general tendency, we obtained smaller uncertainties for longer lived radionuclides than for short-lived ones. In most cases, however, the total uncertainties were less than 10%. All these errors were considered to be independent. Consequently, they were quadratically added according to the laws of error propagation to obtain total errors of the cross sections relative to the standard cross sections. However, some of the sources of errors are common to all data of our experiments in the same way, while others individually affect each reaction. The common errors sum up quadratically and have to be quadratically subtracted when discussing ratios of cross sections from this work or the precision of the measured cross sections instead of their accuracy.

Table (3-7): Major Uncertainty Sources and their Estimated Values in the Cross Section Measurements.

Sources of Uncertainties	Error Magnitude in (%)
Number of Target Nuclei in the Sample (Target mass)	< 1.0
Flux Fluctuation During Irradiation Time	< 1.0
Irradiation, cooling and measuring times	< 1.0
Gamma-gamma coincidence summing	< 1.0
Half-Lives of the radionuclides	1.0
Intensity of Gamma Rays	1.0
Absolute Efficiency Calibration	2.5
Solid angle, γ-attenuation, and γ-self absorption	2.0
Contamination Due to Recoil Losses	Not Considered
Energy derived from stopping power range	Not Considered
Quality of the Monitor Data (Proton Fluence)	5.0
Dead Time and Pile-up Losses in γ-Spectroscopy	< 1.0
Statistical Uncertainty in Net Peak Areas	1.0-5.0
Secondary Particles contribution in Nuclides Production, and Beam Intensity Loss as the Beam Traverses the Stack Thickness	7.0
The Root Average of Total Error Due to all these Factors after Quadrate Sum is Estimated to Be	10.2-11.3

4-1 Introduction

Though several investigations are available in the literature for the determination of reaction cross-sections related to the production of radionuclides, there are large discrepancies in the cross-sections measured for the same reaction by different authors. Further, there are large uncertainties in the measured cross-sections due to the use of low-resolution detectors. Moreover, the details of errors and their evaluation are not discussed in general. Recent experiments have clearly indicated that in statistical nuclear reactions, at moderate excitation energies, particles are emitted prior to the establishment of thermodynamic equilibrium of the compound nucleus (CN). This process is generally known as pre-equilibrium emission (PE).

Through this study measurement and analysis of the excitation functions for a number of reactions using proton beams, the cross-sections for $^{nat}Mo(p,x)^{92m}Nb$, $^{94,95m,95g,96m+g,99m}Tc$, and $^{nat}Gd(p,x)^{152,154m,154g,155,156,160}Tb$ reactions have been measured from threshold up to 18 MeV, using the activation technique. The computer codes ALICE-91 [cf. **29**] is a hybrid code uses the Weisskopf-Ewing evaporation model [cf. **122**], the Bohr-Wheeler model for fission and the geometry dependent hybrid model for precompound decay. Thus it is interesting to compare the excitation functions calculated using this code with the known experimental data. While code EMPIRE-II (M. Herman, 2005) [cf. **30**], this is the modified version of previous release EMPIRE-MSC [cf. **123**], that uses the multistep compound (MSC), and multistep direct (MSD) formulations (Feshbach et al., 1980) [cf. **124**], has been employed for the quantum mechanical calculations. The details of the measurements are presented in next section, while analysis of the data is discussed in next chapter. The excitation functions for the proton induced reactions $^{nat}Gd(p,x)^{154m,154g,155,156,160}Tb$ to the best of our knowledge, have been reported for the first time and hence no comparison with literature data is presented.

Signatures of pre-equilibrium emission are often found in the high energy tails of the excitation functions. The pre-equilibrium emission mechanism has attracted considerable attention from both the experimental and theoretical viewpoints of Gadioli and Hodgson, (1992) [cf. **125**]. The computer codes ALICE-91 Blann, (1971) [cf. **126**-**130**] have been successfully used to describe the experimental data on pre-equilibrium emission. Recently, quantum mechanical (QM) theories have also been used to analyze the experimental data mostly on nucleon induced reactions Feshbach et al., (1980) [cf. **124**,**125**].

4-2 Outline of Nuclear Reaction Mechanisms

In the present work the measured excitation functions for the reactions $^{nat}Mo(p,x)$ and $^{nat}Gd(p,x)$ have been analyzed using both ALICE-91 as well as code EMPIRE-II computer codes with consistent sets of parameters. These parameters for the ALICE-91 approach were obtained from the recent analysis of proton, neutron and alpha induced reactions. The computer codes ALICE-91 Blann et al., (1971, 1972, 1975, 1982, 1988, and 1991; Blann and Vonach, 1983) [cf. **34**, **126**-**130**], the code EMPIRE-II (M. Herman, 2005) [cf. **30**] has been used for description of the data. Brief details of these codes and the parameters used in the calculations are summarized in the following sections. These codes includes models describing the nuclear reaction induced with a projectile have a certain amount of energy through three mechanisms that are the Direct, Pre-equilibrium emission and compound nucleus mechanisms.

4-3 Description of Code ALICE-91

The theoretical cross section calculations of nuclear reactions to predict relevant excitation functions using the ALICE-91 code developed by Blann et al., (1991) [cf. **29**]. The Algorithm of this code depends mainly on evaluation of the emitted nucleon spectra using the following formula:

$$\frac{d\sigma^{pre}}{d\varepsilon_x} = \pi\lambda\!\!\!^{-2}\sum_{l=0}^{\infty}(2l+1)T_l\sum_{n=n_0}^{\pi}R_x(n)\cdot\frac{\omega(p-1,h,E-Q_x-\varepsilon_x)}{\omega(p,h,E)}\cdot\frac{\lambda_c^x}{\lambda_c^x+\lambda_+^x}\cdot g\cdot D_n \qquad (4-1)$$

Where, $\lambda\!\!\!^{-}$ is the reduced wavelength of the incident particle, T_l is the transmission coefficient calculated using optical model, ε_x is the energy of the nucleon emitted, Q_x is the binding energy of nucleon in compound nucleus, ω(p,h,E) is the density of n-exciton states having p particles and h holes (p+h=n) at the excitation energy E, λ_c^x is the intranuclear transition rate corresponding to the absorption of nucleon in nucleus, g is the single particle level density, $R_x(n)$ is the number of x-particles in n-exciton state, D_n is the Depletion factor for n-exciton state, n_0 is the initial exciton number. More details of formulae used to calculate the above parameters are found in [cf. **130**].

The density of exciton states has been calculated with Strutinski-Ericson formula:

$$\omega(p,h,E) = \frac{g(gE-A)^{2-1}}{|p\cdot|h\cdot|(n-1)}, \qquad (4-2)$$

To calculate the emission rate of nucleon, the following relation has been used:

$$\lambda_c^x = \frac{(2s_x + 1) \cdot \mu_x \varepsilon_x \sigma_{inv}^x(\varepsilon_x)}{\pi^2 \cdot g_x^3}, \qquad (4-3)$$

Where, s_x and μ_x are the spin and the reduced mass for particle of x-type, σ_{inv}^x is the inverse reaction cross section for considered particle level density for x-particle.

The internuclear transition rate has been calculated with the following formula:

$$\lambda_+^x = V\sigma_0(\varepsilon_x)\rho_l \qquad (4-5)$$

Where, V is the velocity of nucleon moving inside the nucleus, σ_0 is the Pauli principle corrected cross section of nucleon-nucleon interaction, ρ_l is the nuclear density in the range $l\lambda \prec r_l \prec (l+1)\lambda$.

The factor $R_x(n)$ included in formula (4-1) was calculated for neutron induced reactions by the following way:

$$R_n(3) = \frac{(Z + 2A)}{(2Z + A)};$$
$$R_p(3) = 2 - R_x(3);$$
$$R_x(n) = R_x(3) + \frac{n-3}{4}. \qquad (4-6)$$

The calculation of nonequilibrium spectra was carried out taking into account multiple precompound nucleon. It was assumed that the number of particles emitted from a particular Exciton State is equal:

$$P_{np} = P_n P_p \ (\textit{for emission of neutron and proton});$$
$$P_{nn} = P_n P_n / 2 \ (\textit{for emission of two neutrons}), \qquad (4-7)$$

Here, P_n and P_p are the total number of neutrons and protons emitted from considered n-exciton configuration.

For the code ALICE-91 (release April, 1991), which is a modification from the earlier releases versions of the commonly used hybrid exciton model code ALICE-85/300 (cf. **128**), in the following respects:

(a) Level density options due to Kataria-Ramamurthy [cf. **131**] with inclusion of shell corrections may be selected. Level density options due to Ignatyuk, (1979) (Superfluid Nuclear model) [cf. **132**] may be selected. Some work needs to be done on fine tuning input for this option.

(b) Kalbach systematics C. Kalbach-Cline, (1973) [cf. **133**] for pre-compound angular distributions in as option; it provides much faster algorithm than the option based on n-n scattering kinematics and gives better agreement with experimental results.

(c) Gamma ray spectra, and gamma rays compete with nucleon emission could be produced, this helps to simplfy the problems in case of trapped protons for very proton rich nuclei.

This version may be used to for isotopically mixed targets, e.g. natural isotopic composition. For proton calculations on natural target, the total cross section for the formation of a particular residual isotope was obtained by summing over all fractional isotopic constituents according to the natural abundance of each isotope constitute the natural target. The code has been successfully applied up to 20 MeV to the calculation of (p,xn) reaction cross sections on medium mass target nuclei Lambrecht et al., (1999); Nandy et al., (2001); Hohn et al., (2001) [cf. **134**,**135**,**136**]. However, more complex reactions, like (p,α), were not reproduced well Lambrecht et al., (1999) [cf. **134**]. The ALICE-91 code has been used for the estimation of parameters for the production of medical radioisotopes [cf. **137**-**141**].

With a light projectile (A<5) ALICE-91 describes the interaction potential with the standard optical model code SCAT2 of Bersillon [cf. **142**] in case of spherical nucleus target. The pre-equilibrium mechanism describes the intermediate state between the reactions via a compound nucleus and direct processes. The ALICE-91 code is a hybrid model, the hybrid model is relevant to the precompound decay not to the compound decay. The hybrid means a combination between two theories on the precompound decay. The geometry dependent hybrid model is a further revision of the hybrid model. The statistical model describes the excited states of the nucleons, where Fermi Gas model is applied.

According to this model a particle below the Fermi energy is considered to be in the ground state and the excited particle occupies a level above Fermi energy. The vacant place, which the excited particle leaves below the Fermi level, is called a hole. In advanced approach, a particle-hole system was considered as a quasi particle or exciton [cf. **143**]. From the density of the exciton states and several other parameters it is possible to evaluate the probability of nucleon to be ejected at certain incident particle energy. This calculation results in a particle spectrum over a certain emission energy range. The reaction cross section at a certain incident energy is obtained by integration over produced spectrum.

In the present work, the excitation functions of all ^{nat}Mo(p,x), ^{nat}Gd(p,xn)-reactions were calculated from their respective thresholds up to 20 MeV. The two cases for which the calculations could not be done were the ^{nat}Mo(p,x)^{95m}Tc, ^{nat}Mo(p,x)^{95g}Tc and ^{nat}Gd(p,x)^{154m}Tb, ^{nat}Gd(p,x)^{154g}Tb since ALICE-91 gives only the summed cross section and not the formation of the isomeric states. In all the calculations, standard recommended input data without any additional parameter fitting were used.

4-4 Description of Code EMPIRE-II

In this code theoretical calculations are performed for all reactions contributing to the observed radionuclides, according to Herman et al., (2003) [cf. **30**]. EMPIRE-II is a modular system of nuclear reaction codes, comprising various nuclear models for major reaction mechanisms such as direct, pre-equilibrium, and compound nucleus. If the target is a spherical nucleus, then the interaction with a light projectile (A<5) EMPIRE-II describes the interaction potential, and calculate the transmission coefficients with SCAT2-code the standard Spherical Optical Model of Bersillon [cf. **142**]. EMPIRE-II could be also use the ECIS-code (Coupled Channel Model), and DWBA (Distrorted Wave Born Apploximation) of J. Raynal, (1994) [cf. **144**], especially in the case of deformed targets, Coupled Channel calculations have to be involved in order to obtain adequate transmission coefficients and the low-lying discrete collective level population. The input for ECIS-95 is prepared automatically with parameters retrieved also from RIPL-2 [cf. **145**] or provided by the user.. All calculate the total, elastic and reaction cross sections and provides transmission coefficients for emission of neutrons, protons, α-particles and eventually the ejection of a single light particle (d, t,...). The optical model parameters are retrieved from the RIPL-2 [cf. **145**], prepared internally according to the built-in systematics, or provided by user.

This code originally contained the Hauser-Feshbach theory [cf. **146**], and the classical HYBRID model is simulated employing the geometry dependent hybrid (GDH) model of Blann, (1972) [cf. **127**], to account for the preequilibrium effects. The decay of the compound nucleus is treated either in terms of the Hofmann-Richert-Tepel-Weidenmuller et al., (1975) [cf. **147**], (HRTW) theory accounts for the width fluctuation effects or in terms of standard Hauser-Feshbach (1952) theory [cf. **146**]. The width fluctuation correction was implemented in terms of the (HRTW) approach [cf. **147**], in case open particle channels cause the fluctuations to cancel. The code contains also the multi-step compound mechanism represented in (NVWY) model by Nishioka et al., (1986) [cf. **148**]. This version also included combinatorial calculations of particle-hole level densities. It is designed for calculations over abroad range of energies and incident particles. A comprehensive library of input parameters covers nuclear masses, optical model parameters, ground state deformation, discrete levels and decay schemes, level densities, strength functions, etc.

The reaction cross section is calculated by EMPIRE-II in terms of transmission coefficients $T_l^a(\varepsilon)$ using the expression:

$$\sigma_a(U,J,\pi)=\frac{\pi}{k^2}\frac{(2J+1)}{(2I+1)(2i+1)}\sum_{S=|I-i|}^{J+S}\sum_{l=|J-S|}^{J+S}f(l,\pi)T_l^a(\varepsilon), \qquad (4-8)$$

Where, k is the wave number of the relative motion, i, I, J, and S indicate the spin of the projectile, target, compound nucleus, and the channel spin, respectively, and l is the orbital angular momentum of the projectile a. The function $f(l,\ \pi)$ ensures parity conservation. It is unity if $p*P*(-1)^l=\pi$ and zero otherwise. Here p, P, and π are projectile, target, and compound nucleus parities and ε and U stand for the projectile and compound energy.

In the following section we will demonstrate different models included in the EMPIRE-II code with mathematical interpretation, beside that the code has a graphical user interface which allow an easy exploitation of its capabilities. Graphical representation can be obtained using the plotting package ZVView [cf. **149**].

The program calculates spectra and cross sections for capture and multistep nuclear reactions in the frame of combined pre-equilibrium emission and compound nucleus mechanisms model. Angular momentum is observed throughout the whole calculation and its considerations have been incorporated into the model describing the nucleon emission prior to equilibrium of the excited nucleus. The full γ–cascade, containing transitions from and between the continuum states as well as between the discrete levels, is included in the calculations providing the γ–spectra and the populations of the discrete levels in the residual nuclei.

4-4.1 Multi-Step Direct Model

The approach to statistical Multi-step Direct reactions is based on the Multi-step Direct MSD theory of the pre-equilibrium scattering to the continuum, originally M. Chadwick [cf. **150**] through the code (ORION & TRISTAN). This approach has been revised, especially the part related to statistical and dynamical treatment of nuclear structure. The evolution of the projectile-target system from small to large energy losses in the open channel space is described in the MSD theory with a combination of direct reaction DR, microscopic nuclear srtucture, and statistical methods. As typical for the DR-approach, it is assumed that the closed channel space, i.e. the MSC contributions, have been projected out and can be treated separately within the Multi-step Compound mechanism.

Second-chance preequilibrium Emission after MSD has been incorporated in connection with the MSD model. A semi-classical formulation proposed by M. Chadwick [cf. **150**] has been adopted. The current implementation assumes that MSD emission leaves residual nucleus containing a single particle-hole pair of energy equal to the excitation energy of the residual. The excited particle (neutron or proton) is given a chance to escape the nucleus. The emission rate is defined by the product of the related s-wave transmission coefficient and the probability for the exciton to have excitation energy compatible with the channel energy. Angular momentum coupling assumes lp-lh spin distribution for the exciton pair. This treatment allows accounting only for the second chance emission. Thus it extends applicability of the MSD model up to the energies at which higher order emissions can still be neglected. In general it results in a modest increase of the central part of the emission spectra. The final formula for the two step MSD cross section is:

$$d^2\sigma^2 \Big/ dE\, d\Omega = \sum_{\lambda_1\lambda_2} \int dE_1\, S_{\lambda_2}(E,E_1)\, S_{\lambda 1}(E,0)\, \overline{\frac{d\sigma^{(2)}}{d\Omega}}(E,E_1)\Big|_{\lambda_1\lambda_2}. \qquad (4-9)$$

$\overline{\frac{d\sigma^{(2)}}{d\Omega}}$, is an averaged cross section,

$$S_{\lambda}(E,E_1)=\frac{\sum_{c_1}P_{c_1}(E_1)S_{\lambda}(E,c_1)}{\sum_{c_1}P_{c_1}(E_1)}, \qquad (4-11)$$

$S_{\lambda}(E,c)$ is the transition strength function. Theoretically, good approximation of $S_{\lambda}(E, c)$ could be achieved by taking the average over the response functions belonging to c_1 at energy E_1:

$$P_c(E)=-\frac{1}{\pi}Im\left[\int dE' g(E-E')\left(c\middle|G^{\text{int}r}(E')\middle|c\right)\right], \qquad (4-10)$$

$P_c(E)$ is the probability per energy to find the system in the configuration c, is given by the spectroscopic densities. Integrating $P_c(E)$ over an interval ΔE, we obtain the spectroscopic factor for the configuration c in ΔE.

In the present version of the code only one and two step MSD contribution are considered.

4-4.2 Multi-Step Compound Model

The modeling of Multi-step Compound (MSC) processes follows the (NVWY) approach of Nishioka et al., (1986) [cf. **148**]. Like most of the pre-compound models, the (NVWY) theory describes the equilibration of the composite nucleus as a series of transitions along the chain of classes of closed channels of increasing complexity. In the present context, we define the classes in terms of the number of excited particle-hole pairs (n) plus the incoming nucleon, i.e. excitons. Thus the exciton number is $N = 2n + 1$ for nucleon induced reactions. Assuming that the residual interaction is a two-body force only neighboring classes are coupled (An = ±1).

According to NVWY, the average MSC cross-section leading from the incident channel *a* to the exit channel *b* is given by:

$$\frac{d\sigma_{ab}}{dE}=(1+\delta_{ab})\sum_{n,m}T_n^a\prod\nolimits_{n,m}T_n^a, \qquad (4-12)$$

Which also has to be summed over spins and parities of the intermediate states and where we have omitted kinematic and angular-momentum dependent factors. The summation includes all classes *n* and *m*. The transmission coefficients T_n^a describing the coupling between channel *a* and class *n* are given as:

$$T_n^a = \frac{4\pi^2 U_n^a}{\left(1+\pi^2 \sum_m U_m^a\right)^2}, \qquad (4-13)$$

where $U_n^a = \rho_n^b \langle W_{n,a} \rangle$ is microscopically defined in terms of the average bound level density ρ_n^a of class n, and in terms of the average matrix elements $W_{n,a}$ connecting channel a with the states in class n. The probability transport matrix Π_{mn} is defined via its inverse:

$$\left(\Pi^{-1}\right)_{nm} = \delta_{nm}\left(2\pi\rho_n^b\right)\left(\Gamma_n^{\downarrow} + \Gamma_n^{ext}\right) - \left(1-\delta_{nm}\right)2\pi\rho_n^b \overline{V_{n,m}^2}\, 2\pi\rho_m^b, \qquad (4-14)$$

in terms of the mean squared matrix element $\overline{V_{m,n}^2}$ coupling states in classes n and m, the average spreading width $\Gamma_n^{\downarrow}$ of states in class n, and the average total decay width Γ_n^{ext} in class n. The spreading width $\Gamma_n^{\downarrow}$ is again related to the mean squared matrix element $\overline{V_{m,n}^2}$ as follow:

$$\Gamma_n^{\downarrow} = 2\pi \sum_m \overline{V_{n,m}^2}\, \rho_m^b. \qquad (4-15)$$

Under the chaining hypothesis $\overline{V_{m,n}^2}$ couples only neighboring classes ($\overline{V_{m,n}^2}$ = 0 unless $|n—m| = 1$). The decay width Γ_n^{ext} is determined by the sum of the transmission coefficients T_n^a over all open channels:

$$\Gamma_n^{ext} = \left(2\pi\rho_n^b\right)^{-1} \sum_a T_n^a, \qquad (4-16)$$

More explicitly Γ_n^{ext} may be expressed through the energy integral of the product of transmission coefficients and level densities:

$$\Gamma_n^{ext} = \left(2\pi\rho_n^b\right)^{-1} \sum_a \sum_{m=n-1}^{m=n+1} \int T_{n+m}^a(\varepsilon)\rho_m^b\left(E - Q_p - \varepsilon\right)d\varepsilon. \qquad (4-17)$$

Here, ε stands for the ejectile p energy, Q_p for its binding in a composite system, and a symbolically accounts for the angular momentum coupling of the residual nucleus spin, ejectile spin and orbital angular momentum to the composite nucleus spin. Again, due to the chaining hypothesis, only those emissions which change class number by $|n—m| < 1$ are allowed.

We note that, in the NVWY theory, the transmission coefficients $T_{n\rightarrow m}$ carry two class indices.

4-4.3 Coupling Between MSC, and MSD

The NVWY theory includes a possibility of feeding higher MSC classes directly from the MSD chain, in addition to the normal transitions between bound states of increasing complexity. This process is taken into account by a double sum over classes in the cross section formula Eq. (4-18). The second sum over n refers to the contribution of different classes to the particle emission, while the first one (over m) corresponds precisely to the population of various classes directly from the open channel space rather than through the transitions along the MSC chain. This effect is included in the EMPIRE-11 code by distributing the incoming channel transmission coefficient over different MSC classes. It is done according to phase space, and global coupling arguments requiring that the incoming flux splits between the first MSD and MSC classes in proportion to the respective state densities, and to the average value of the squared matrix elements coupling unbound to unbound $\left(\left\langle V_{uu}^2\right\rangle\right)$ and unbound to bound states $\left(\left\langle V_{ub}^2\right\rangle\right)$. Introducing $R=\left\langle V_{uu}^2\right\rangle\left|\left\langle V_{uu}^2\right\rangle\right.$, denoting the optical model transmission coefficient by Tom, the density of bound and unbound states in class n by ρ_n^b and ρ_n^u respectively, and their sum by p, the transmission coefficient populating the first MSC class may be written as:

$$T_1 = T_{om}\frac{\left\langle V_{ub}^2\right\rangle \rho_1^b(E)}{\left\langle V_{ub}^2\right\rangle \rho_1^b(E)+\left\langle V_{uu}^2\right\rangle \rho_1^u(E)} = T_{om}\frac{R}{(R-1)+\frac{\rho_1(E)}{\rho_1^b(E)}}, \qquad (4-18)$$

The same reasoning may be applied to the flux remaining in the open space, which may enter the MSC chain in subsequent steps of the reaction. Assuming R to be independent of the class number, the transmission coefficient T_n populating the nth MSC class is written as:

$$T_m = T_{om} - \sum_{i=1}^{n-1} T_i \frac{R}{(R-1)+\frac{\rho_n}{\rho_n^b}}, \qquad (4-19)$$

If the MSD option is- selected the absorption cross section available to MSC *(σ_{abs})* is reduced by the total MSD emission cross section *(σ_{MSD})* m order to ensure flux conservation and becomes:

$$\sigma_{abs}(J)=\sigma_{OM}(J)\left(1-\frac{\sigma_{MSD}}{\sigma_{OM}}\right), \qquad (4-20)$$

Where, *(σ_{OM})* is optical model reaction cross section and *J* stands for the compound nucleus spin.

4-4.4 Exciton Model

This module was incorporated to improve (n, γ), and (p, γ) reactions for fast neutrons and to add the capability of predicting spectra for the charge-exchange reactions that are not provided by the current implementation of the MSD model (ORION & TRISTAN). DEGAS is the exciton model code with angular-momentum conservation written by E. Betak [cf. **151**]. DEGAS is the exciton model code with angular-momentum conservation. Inclusion of the Monte Carlo Preequilibrium Emission (DDHMS) by M.B. Chadwick [cf. **150**] extends useful energy range of EMPIRE-II up to the intended limit of about 200 MeV for nucleon induced reactions. The HMS model, developed by M. Blann [cf. **152**] is an exciton model inspired treatment of the intranuclear cascade. Certain features of the DEGAS code were intentionally disabled for compatibility with other models present in the EMPIRE-II code. Thus, γ-cascade has been limited to primary γs in the composite nucleus and multiple pre-equilibrium emissions have been blocked. In particular, DEGAS treatment of the equilibrium emission was disabled and left to the Hauser-Feshbach model coded in EMPIRE. Within the above simplifications DEGAS solves the classical (apart of spin) set of master equations:

$$\begin{aligned}\frac{dP(E,J,n,t)}{dt} &= P(E,J,n-2,t)\lambda^{+}(E,J,n-2)+P(E,J,n+2,t)\lambda^{+}(E,J,n+2)\\ &+P(E,J,n,t)\left[\lambda^{+}(E,J,n)+\lambda^{-}(E,J,n)+L(E,J,n)\right]\\ &+\sum_{J',n',x}\int P(E',J',n',t')\lambda_{x}\left(\left[E',J',n'\right]\rightarrow\left[E,J,n\right]\right)d\varepsilon, \qquad (4-21)\end{aligned}$$

Where, *P(E,J,n,t)* is the occupation probability of the composite nucleus at the excitation energy *E*, spin *J* and the exciton number n, λ^{+} and λ^{-} are the transition rates for decay to neighboring states, and *L* is the total integrated emission rate for particles protons, neutrons, and γ-rays. Note that the last term ensures coupling of different spins.

The nucleon emission rate per energy and time is defined as:

$$\lambda_{\pi,\nu}\left(\left[E,J,n\right]\rightarrow\left[U,S,n-1\right]\right)=\frac{1}{h}\frac{\omega\left(n-1,U,S\right)}{\omega\left(n,U,S\right)}\Re_{\pi\nu}\left(n\right)\sum_{L=\left|S-\frac{1}{2}\right|l=\left|J-j\right|}^{J-j}T_{l}\left(\varepsilon\right), \qquad (4-22)$$

Where, ω(n, E, J) is the particle-hole state density, $T_{l,S}$ are the transmission coefficients of the emitted nucleon, and $\Re_x(n)$ is a fraction of x-type nucleons in the n-th stage. The particle-hole state density is:

$$\omega\left(n,E,J\right)=\frac{g\left(gE-A_{p,h}\right)}{p\left|h\right|\left(n-1\right)\left|\right.}R_{n}\left(J\right), \qquad (4-23)$$

Where, g is the single-particle level density, p and h are the number of particles and holes (n = p + h), and $A_{p,h}$ is the Pauli correction term. The spin distribution reads:

$$R_{n}\left(J\right)=\frac{2J+1}{2}\exp\left(-\frac{\left(J+\frac{1}{2}\right)^{2}}{2\sigma_{n}^{2}}\right), \qquad (4-24)$$

With σ_n being the spin cut-off parameter $\left(\sigma_n^2=\left(0.24+0.0038E\right)nA^{2/3}\right)$. The energy and spin dependence are assumed to factorized as follow:

$$\lambda^{\pm}\left(E,J,n\right)=\frac{2\pi}{\hbar}\left|M\right|^{2}Y_{n}^{\downarrow}X_{nJ}^{\downarrow}, \qquad (4-25)$$

Where, $\left|M\right|^2$ is the energy part of the average squared transition matrix element of the residual interaction, $Y_n^{\downarrow}$ is the energy part of the density of accessible final state, and $X_{n,J}^{\downarrow}$ factor take care of angular momentum coupling.

DEGAS solves the set of master equations Eq. (4-21) and calculate the integrals of occupation probabilities:

$$\tau(E,J,n)=\int_0^{\infty}P(E,J,n,t)\,dt, \qquad (4-26)$$

DEGAS pre-equilibrium emission cross sections is given by:

$$\frac{d\sigma_x}{d\varepsilon_x}=\sum_{J_c,J_r,n}\int\sigma(E_c,J_c)\tau(E_c,J_c,n)\lambda_x\left(|E_c,J_c,n|\rightarrow|E_r,J_r,n-1|\right)dE_r. \qquad (4-27)$$

Here c and r subscriptions refer to the composite and residual nuclei respectively.

4-4.5 Monte Carlo Preequilibrium Model

The intermediate state between the reactions via a compound nucleus and direct processes is described by the pre-equilibrium mechanism. Inclusion of the Exciton Model DEGAS was incorporated to add the capability of predicting spectra for charge exchange reaction that is not provided by the current implementation of the MSD model (ORION&TRISTAN codes). DEGAS is the exciton model code with angular-momentum conservation. E. Betak (1993) [cf. **151**]. Inclusion of the Monte Carlo Preequilibrium Emission (DDHMS) by M.B. Chadwick [cf. **150**] extends useful energy range of EMPIRE-II up to the intended limit of about 200 MeV for nucleon induced reactions.

The Hybrid Monte-Carlo Simulation (HMS) approach to the preequilibrium emission of nucleons is the third precompound model included in EMPIRE-11 (the other two are (MSD)&(MSC) and the exciton model (DEGAS)). M. Blann [cf. **126**] has formulated, the original HMS model as a hybrid model inspired version of the intranuclear cascade approach. Contrary to other classical preequilibrium models, this approach avoids multi-exciton level densities to be used inconsistently n he xciton, and in the hybrid formulations. The HMS model, developed by M. Blann [cf. **152**] is an exciton model inspired treatment of the intranuclear cascade. Each exciton is allowed to create a new particle-hole pair or escape the composite nucleus if the exciton is of a particle (rather than hole) type.

The Monte Carlo method is used to decide whether the interaction counterpart is a neutron or proton, to select particle-hole pair energy and to choose exciton directions. The cascade is terminated when all excitons fall below binding energy. The HMS model needs only lp-lh level densities, therefore it avoids inconsistent use of multi-exciton level densities typical of classical preequilibrium models. An important feature of the model is that it takes into account unlimited number of multi-chance preequilibrium emissions and observes angular momentum and linear momentum coupling. It provides double differential emission spectra of neutrons and protons, the populations of residuals depend on its spin and its excitation energy, as well as the recoil spectra depends on the excitation energy too.

These results are transferred onto EMPIRE-II arrays and used as a starting point for the subsequent compound nucleus decay. The HMS model has a number of attractive features. First of all, there are no physical limits on a number of preequilibrium emissions (apart from energy conservation). With the addition of linear momentum conservation by M. Chadwick and P. Oblozinsky (DDHMS) [cf. **150**], the model provides a nearly complete set of observables. Spin and excitation-energy dependent populations of residual nuclei can also be obtained an essential feature for couplng the pre-equilibrium mechanism to be the subsequent Compound Nucleus decay.

The binding energies in the HMS model are thermodynamically correct, this is clear improvement over the intranuclear cascade model, although of the exact account, typical of the Compound Nucleus model, is still out of reach. The DDHMS model proved to perform very well up to at least 250 MeV, extending energy range of the EMPIRE-11 applicability to the desired limit.

The calculation flow in the DDHMS model can be summarized in terms of the following steps:

1. Draw collision partner for the incoming nucleon (2p-lh state created)
2. Draw energy (c) of the scattered nucleon (if bound go to step 5)
3. Draw scattering angles for both particles
4. Decide whether the-scattered nucleon will be emitted, re-scattered or trapped,
   a) if emitted appropriate cross section is augmented
   b) if re-scatters additional particle-hole is created and we return to step 2
   c) if trapped, go to step 5
5. Draw excitation energy of a particle in the remaining Ip-lh configuration (between 0: (U- ε), if unbound go to step 3, if bound choose another existing Ip-lh pair and repeat step 5.

All excitons (including holes) are treated on equal footing and each of them is given a chance to interact or being emitted in priority with an equal probability. The cascade ends when all excitons are bound. Below, we summarize various probability distributions that are used in concert with a random number generator.

For choosing a collision partner it is assumed that the unlike interaction is 3 times more probable than the like one ($\sigma_{np} = 3\sigma_{nn}$). Thus, for the incident neutron we have P_{nn} and P_{np} for the probability of exciting neutron and proton respectively, defined as:

$$P_{nn} = \frac{(A-Z)}{(A-Z)+3Z'}, \qquad (4-28)$$

$$P_{np} = 1 - P_{nn}, \qquad (4-29)$$

Similarly for the incident proton is:

$$P_{pp} = \frac{Z}{Z+3(A-Z)}, \qquad (4-30)$$

$$P_{pn} = 1 - P_{pp}, \qquad (4-31)$$

The energy distribution of the scattered particles P(ε) is given by the ratio of the (n-1), and n-exciton level densities ρ_n:

$$P(\varepsilon)d\varepsilon = \frac{\rho_{n-1}(E-\varepsilon)g}{\rho_n(E)}, \qquad (3-32)$$

The emission probability is calculated as:

$$P_v(\varepsilon - Q) = \frac{\lambda_c(\varepsilon-Q)}{\lambda_c(\varepsilon-Q)+\lambda_+(\varepsilon)}, \qquad (4-33)$$

The emission rate being:

$$\lambda_c(\varepsilon-Q) \sim \frac{\sigma_v(\varepsilon-Q)(\varepsilon-Q)(2S+1)\mu_v}{g}. \qquad (4-34)$$

σ_v, is an inverse reaction cross section, Q is a binding energy, g is a single particle density, S denotes nucleon spin, and μ_v stands for the reduced nucleon mass. Following the hybrid model, $\lambda_+(\varepsilon)$ is calculated from the mean free path of a nucleon in nuclear matter.

4-4.6 Compound Nucleus Model

The statistical compound nucleus model in the Hauser-Feshbach formalism [cf. **146**] is implemented in EMPIRE-II. Exact angular moment and parity coupling is considered. The emission of neutrons, protons, and α-particles is taken into account in competition with full γ-cascade in the residual nuclei. When the compound nucleus is populated at low excitation energies, elastic channel enhancement and width fluctuations must be taken into account. In EMPIRE-II these effects are considered in the frame of HRTW model [cf. **147**].

The main quantities entering the Hauser-Feshbach formalism for reaction cross section calculations of non-fissioning nuclei are the transmission coefficients for particles and gamma decay. The particle transmission coefficients are provided by SCAT2 or ECIS-95 subroutines. The gamma-transmission coefficients are calculated taking into account E1, E2 and M1 multipolarities. An arbitrary mixture of Weisskopf single particle model and giant multipole resonance of Brink-Axel [cf. **153**] can be used.

After particle or gamma ray emission the nuclei can be populated in discrete or continuum states. Therefore, the transmission coefficients calculation requires the knowledge of the discrete level schemes and of the density functions describing the continuum part of the excitation spectra for all the nuclei involved in reaction. The level densities are described in EMPIRE-II by different models and parameterizations. These may be Gilbert-Cameron, (1965) [cf. **154**] type formulae (with constant or energy-dependent constant level density parameter a) or dynamical approaches which taken into account collective enhancements [cf. **30**]. For each option, the parameters can be retrieved from built-in systematics, from RIPL-2 [cf. **145**], or provided by the user.

The statistical compound nucleus model in the Hauser-Feshbach formalism [cf. **146**] is implemented in EMPIRE-II. The exact angular momentum and parity coupling is observed. The emission of neutrons, protons, α-particles, and a light ion is taken into account along with the competing fission channel. The full γ-cascade in the residual nuclei is considered. Particular attention is dedicated to the determination of the level densities, which can be calculated in the non-adiabatic approach allowing for the rotational and vibrational enhancements. These collective effects are gradually removed above certain energy. Level densities acquire dynamic features through the dependence of the rotational enhancement on the shape of a nucleus.

The particle transmission coefficients are provided by SCAT2 or ECIS-95 subroutines. The gamma-transmission coefficients are calculated taking into account E1, E2 and M1 multipolarities. An arbitrary mixture of Weisskopf single particle model and giant multipole resonance of Brink-Axel [cf. **153**] can be used.

In the frame of the statistical model of nuclear reactions the contribution of the Compound Nucleus (CN) satate a with spin J, parity π, and excitation energy E to a channel b is given by the ratio of channel width Γ_b, to the total width $\Gamma_{tot} = \Sigma_c \Gamma_c$ multiplied by the population of this state $\sigma_a(E,J,\pi)$. This also holds for secondary CNs that are formed due to subsequent emissions of particles. The only difference is that while the first CN is initially excited to the unique (incident channel compatible) energy, the secondary CNs are created with excitation energies which spread over the available energy interval. Each such state contributes to the cross section by:

$$\sigma_b(E,J,\pi) = \sigma_a(E,J,\pi)\frac{\Gamma_b(E,J,\pi)}{\sum_c \Gamma_c(E,J,\pi)}, \qquad (4-35)$$

4-4.7 Level Densities

EMPIRE-11 accounts for various models describing level densities and includes several respective parameterizations. In each case equal parity distribution $\rho(E, J, \pi) = ½\ \rho(E, J, \pi)$ is assumed. Choice of the proper representation depends on a case being considered. For the nucleon-induced reactions, with CN excited up to about 20 MeV, the Gilbert-Cameron approach is recommended. It assures the most accurate description of level densities in the energy range up to the neutron binding energy. The collective effects are included in the level density parameter a, providing reasonable estimate of the level densities as long as damping of the collective effects is irrelevant. The relatively low angular momentum introduced by the incident projectile justifies neglect of dynamical effects. However, these effects have to be taken into account in case of the Heavy Ion induced reactions and/or higher excitation energies. In these cases, the dynamic approach that accounts for the shape-dependent collective enhancements, their damping with increasing energy, and the temperature dependence of the a-parameter should be adopted.

4-4.7.1 Gilbert-Cameron Approach

The Gilbert-Cameron approach [cf. **154**] splits excitation energy in two regions. Different functional forms of level densities are applied in each of them. At low excitation energies (below the matching point U_x) the constant temperature formula is used:

$$\rho_T(E) = \frac{1}{T} exp\left[(E - \Delta - E_0) \big/ T \right], \qquad (4-36)$$

Where T is the nuclear temperature, E is the excitation energy ($E = U + \Delta$ with Δ being the pairing correction), and E_0 is an adjustable energy shift. Above U_x the Fermi gas formula is applied:

$$\rho_F(U) = \frac{exp\left(2\sqrt{aU}\right)}{12\sqrt{a\sigma(U)}\,a^{1/4}U^{5/4}}, \qquad (4-37)$$

The level density parameter a is assumed to be energy independent. The spin cut-off factor $\sigma(U)$ is given by:

$$\sigma^2(U) = 0.146\,A^{2/3}\sqrt{aU}, \qquad (4-38)$$

Three model parameters, T, U_x, and E_0, are determined by the requirement that the level density and its derivative are continuous at the matching point U_x, and by fitting cumulative number of discrete levels with the integral of Eq. (4-36). The first of the conditions implies:

$$\frac{1}{T} = \sqrt{a/U_x} - \frac{3}{2U_x}, \qquad (4-39)$$

The code will not stop if the discrete level scheme is incompatible with the level density parameter a (i.e., there is no real solution to Eq. (4-39). It is essential that only complete level schemes be used for the determination of T and E_0 parameters. With increase in the excitation energy, some levels could be considered in the calculations, numbers of the levels that are supposed to form a complete set is read by EMPIRE-II (starting with version 2.17) directly from the file containing discrete levels *(empire/'data/RIPL /levels/Z*.DAT)*. This database makes part of the RIPL-2 library [cf. **145**].

Choosing the FITLEV option in the input can inspect cumulative plots of discrete levels along with the constant temperature fits. For each nucleus involved in the calculations an appropriate plot is created and stored in the *CUMULPLOT.PS (*-cum.ps)* file, which can be viewed once the calculations are done (NOTE: this option requires "gnuplot" packet and "ps2ps" which is a part of the "ghostscript").

Calculations can be performed using constant level density parameter a read from the input file. No energy dependence is allowed in such a case. Alternatively, one of the built in systematics can be used. The latter ones account for the shell effects, which fade-out with increasing energy implying energy dependence of the *(a)* parameter. The general form of this dependence was proposed by Ignatyuk et al., (1979) [cf. **132**].

$$a(U)=\tilde{a}\left[1+f(U)\frac{\delta W}{U}\right], \qquad (4-40)$$

Where δW is the shell correction, ã is the asymptotic value of the a-parameter, and $f(U)$ is given by:

$$f(U)=1-exp(-\gamma U), \qquad (4-41)$$

Where, γ is a systematic constant, we stress again that Gilbert-Cameron approach does not account explicitly for the collective enhancements of the level densities. These are included implicitly in the (a) when fitting neutron resonance spacings. Such an approach leads to the over-estimation of the level densities above 20 MeV. This deficiency can be overcome using the dynamic approach described below.

4-4.7.2 Dynamic Approach

The dynamic approach to the level densities is specific to the EMPIRE-II code. It takes into account collective enhancements of the level densities due to nuclear vibration and rotation. The formalism uses the super-fluid model below critical excitation energy (when the EMPIRE-II specific parameterization of the level density parameter is selected) and the Fermi gas model above. Differently from other similar formulations, the latter one accounts explicitly for the rotation induced deformation of the nucleus, which becomes spin dependent. The deformation enters level densities formulas through moments of inertia and through the level density parameter (a) that increases with increase in the surface of the nucleus.

Assuming that the prolate nuclei rotate along the axis perpendicular to the symmetry axis the explicit level density formulas reads:

$$\rho(E,J,\pi)=\frac{1}{16\sqrt{6\pi}}\left(\frac{\hbar^2}{\Im_{\parallel}}\right)^{1/2} a^{1/4}\sum_{K=-J}^{J}\left(U-\frac{\hbar^2K^2}{2\Im_{eff}}\right)^{-5/4}$$

$$exp\left\{2\left[a\left(U-\frac{\hbar^2K^2}{2\Im_{eff}}\right)\right]^{1/2}\right\}, \qquad (4-42)$$

In the case of the oblate nuclei which are assumed to rotate parallel to the symmetry axis we have:

$$\rho(E,J,\pi)=\frac{1}{16\sqrt{6\pi}}\left(\frac{\hbar^2}{\Im_{\parallel}}\right)^{1/2} a^{1/4}\sum_{K=-J}^{J}\left(U-\frac{\hbar^2\left[J(J+1)-K^2\right]}{2\left|\Im_{eff}\right|}\right)^{-5/4}$$

$$exp\left\{2\left[a\left(U-\frac{\hbar^2\left[J(J+1)-K^2\right]}{2\left|\Im_{eff}\right|}\right)\right]^{1/2}\right\}, \qquad (4-43)$$

a is a level density parameter, J is a nucleus spin and K its projection, E is the excitation energy and U *is* the excitation energy less pairing (A).

4-4.7.3 Hartree-Fock-BCS Approach

EMPIRE can read precalcuated level densities from the RIPL-2 library, which contains tables of level densities [cf. **145**] for more than 8000 nuclei calculated in the frame of the Hartree-Fock-BCS approach, is the generalized Hartree-Fock theory presented by John Bardeen, Leon Cooper, and Robert Schieffer (BCS). These microscopic results include a consistent treatment of shell corrections, pairing correlations, deformation effects and rotational enhancement. The results were re-normalized to the experimental s-wave neutron resonance spacings and adjusted to the cumulative number of discrete levels, so that the degree of accuracy is comparable to the phenomenological formulae.

Using the partition function method, the state density can be obtained as:

$$\omega(U)=\frac{exp\left[S(U)\right]}{(2\pi)^{3/2}\sqrt{Det(U)}}, \qquad (4-44)$$

Entropy S, and excitation energy U are drived from the summation over single particle levels, and Det stands for the determinant. Pairing correlations are treated within the standard BCS theory in the constant G approximation with blocking. Consequently single particle energies are replaced by their quasi-particle equivalents with BCS equations determining gap parameter Δ and the chemical potential λ.

Spherical and deformed nuclei are treated in a distinct mode. The level density for spherical nuclei is simply related to the state density:

$$\rho_{sph}(U,J) = \frac{2J+1}{2\sqrt{2\pi^3}} \cdot exp\left[\frac{-J(J+1)}{2\sigma^2}\right].\omega(U), \qquad (4-45)$$

While for deformed nuclei the formula is similar to the one used in the EMPIRE-specific level densities:

$$\rho_{def}(U,J) = \frac{1}{2}\sum_{K=-J}^{J}\frac{1}{\sqrt{2\pi\sigma^2}}.exp\left\{-\left[\frac{J(J+1)}{2\sigma_{\perp}^2}+\frac{K^2}{2}\left(\frac{1}{\sigma^2}-\frac{1}{\sigma_{\perp}^2}\right)\right]\right\}.\omega(U), \qquad (4-46)$$

Where, σ is the spin cut-off parameter and $\sigma_{\perp}$ the perpendicular spin cut-off parameter, both affected by the pairing correlations, and $\sigma_{\perp}$ being related to the perpendicular moment of inertia.

4-4.8 Width Fluctuation Correction

The Hauser-Feshbach model assumes that there are no correlations among various reaction channels contributing to the formation and decay of a certain compound nucleus state J^{π}. An obvious case for which this assumption breaks down is the elastic channel, for which entrance and exit channels are the same. This correlation leads to an increase of the elastic channel cross section compared to the Hauser-Feshbach predictions. At higher incident energies, at which the elastic channel is one among many inelastic ones, this increase can be safely neglected. However, if there are only a few open channels the increase of the elastic channels turns out to have serious consequences on other channels (particularly on inelastic scattering and capture).

Hofmann, Richert, Tepel and Weidenmueller (HRTW) [cf. **147**] have proposed one of the models, which allow these effects to be taken into account. In the case of no direct reaction contribution, the averaged S-matrix element connecting channels a and b can be written as:

$$\langle S\rangle_{ab} = \delta_{ab} \cdot exp\left(i S_{ab}\right) \cdot \left(1 - T_a\right)^{1/2}, \qquad (4-47)$$

Where,

$$T_a = 1 - \left|\langle S\rangle_{aa}\right|^2, \qquad (4-48)$$

T_a is an optical model transmission coefficient. The HRTW model assumes that the Compound Nucleus (CN) cross sections factorize and can be expressed through a product of the channel dependent quantities ξ. This would be the famous Bohr's assumption if not for the elastic enhancement factor W_a, which has been introduced by HRTW in order to account for the elastic channel correlation as follow:

$$\left\langle \sigma_{ab}^{fl} \right\rangle = \xi_a \xi_b \quad a \neq b \quad and \quad \left\langle \sigma_a^{fl} \right\rangle = W_a \xi_a, \qquad (4-49)$$

Setting

$$\xi_a = \frac{V_a}{\sqrt{\sum_c V_c}}, \qquad (4-50)$$

We get for the Compound Nucleus (CN) cross section as:

$$\sigma_{ab}^{CN} = \left\langle \sigma_{ab}^{fl} \right\rangle = V_a V_b \left(\sum_c V_c \right)^{-1} \left[1 + \delta_{ab}\left(W_a - 1\right)\right], \qquad (4-51)$$

Taking into account that the incoming flux has to be conserved (unitarity condition) we find the relation between V_S, the elastic enhancement factor (W_a), and the transmission coefficient (T_a):

$$V_a = T_a \left[1 + \frac{V_a}{\left(\sum_c V_c\right)} \left(W_a - 1\right) \right]^{-1}, \qquad (4-52)$$

This equation can be solved for V_a by iteration once all W_a are known. The current version of EMPIRE-11 uses W_a derived from the analysis of numerically generated sets of s-matrices. This exercises was tailored to the cases that include many weak channels coupled to a few strong channels, which is typical of neutron capture reactions. The resulting formula for the elastic enhancement factor is:

$$W_a = 1 + 2\left[1 + T_a^F\right]^{-1} + 87\left(\frac{T_a - T_{ave}}{\sum_c T_c}\right)^2 \left(\frac{T_a}{\sum_c T_c}\right)^5, \qquad (4-53)$$

With

$$F = 4\frac{T_{ave}}{\sum_c T_c}\left(1 + \frac{T_a}{\sum_c T_c}\right)\left(1 + 3\frac{T_{ave}}{\sum_c T_c}\right)^{-1}, \qquad (4-54)$$

Which completes formulation of the model.

4-4.9 Flow of EMPIRE-II Calculations

Listed below are the essential steps in the execution of the general case involving MSD and MSC calculations:

1. Read EMPIRE-II input file (.inp),

2. Construct table of nuclei involved,

3. Read from the input parameter library (or input/output files if they exist),
   a) Discrete levels,
   b) Binding energies,
   c) Level density parameters,
   d) Shell corrections,
   e) Ground state deformations,

4. Calculate:
   a) Transmission coefficients,
   b) Level densities,
   c) Fission barriers

5. Retrieves experimental data from the EXFOR library.

6. Writes input/output files:
   a) *-exf,
   b) *-omp.int,

c) *-omp.dir,
d) *-omp.ripl,
e) *.lev,
f) *-lev.col,

7. Determine fusion cross section,

8. Select the compound nucleus for consideration,

9. Calculate double-differential cross sections for inelastic scattering in terms of the MSD mechanism, populate residual nucleus continuum and discrete levels, store recoil spectra (first CN only),

10. Calculate second-chance preequilibrium emission following MSD mechanism, store spectra of particles and recoils and increment residuals populations. Although MSD is limited to a single type of nucleon, both neutron and protons are emitted in the second-chance process (first CN only),

11. Calculate neutron, proton, and γ emission spectra in terms of the exciton model (code DEGAS), populate residual nuclei continuum and discrete levels (first CN only),
12. Calculate neutron, proton, and γ emission spectra in terms of the HMS model (code DDHMS), populate residual nuclei continuum and discrete levels, store recoil spectra (first CN only),
13. Calculate neutron, proton, and γ emission spectra in terms of the MSC mechanism, populate residual nuclei continuum and discrete levels (first CN only),

14. Calculate neutron, proton, α, γ, and eventually light ion emission widths in the frame of the Hauser-Feshbach model,

15. Calculate fission width in the frame of the Hauser-Feshbach model,

16. Normalize emission and fission widths with the Hauser-Feshbach denominator and fusion cross section to obtain compound nucleus spectra and population of continuum and discrete levels in residual nuclei,

17. Save, print results for the decay of the nucleus considered,

18. Select new nucleus and repeat steps 14 through 18 until all requested nuclei have been processed. The selection scheme is as follows:

Starting from the compound nucleus, neutrons are subtracted until the number of neutron emissions specified in the input is reached. Then one proton is subtracted from the compound nucleus, and all nuclei with decreasing neutron number are considered again. In the Z-N plane the calculations are performed row-wise from the top-right corner (compound nucleus) to the left. When a row is completed the one below is considered,

19. Save, print inclusive spectra, read new incident energy from the input file (*.inp) and repeat steps 4 and 7 through 19.

4-4.10 RIPL-2 Input Parameters Libraries

When not provided by the user, EMPIRE-II retrieves the input parameters either from internal systematics or from different segments of RIPL-2 (Reference Input Parameters Library, version 2) [cf. **145**]. The segments of interest for this work are masses, levels and optical model parameters. The masses file contains the most recent experimental masses recommended by Audi and Wapstra, (1995) [cf. **155**] and the calculations of Moller and Nix, (1981) [cf. **156**]. Besides masses, the file contains the nuclear deformations in the ground state expressed in different parameterizations.

The Levels file contains more than 2500 level schemes originating from ENSDF 1998. For the purpose of reaction calculations the basic ENSDF sets were completed the spins and the parities up to certain cut-off energies by Budapesta group. Details concerning these assignments are published in RIPL-2 documentation [cf. **145**]. The optical model parameters file includes 409 sets of optical parameters for nuclei up to Z=103 for neutrons, protons, deuterons, tritons, ^{3}He and α-particles. EMPIRE-II contains also optical parameters for ^{6}Li, ^{7}Li, and ^{7}Be.

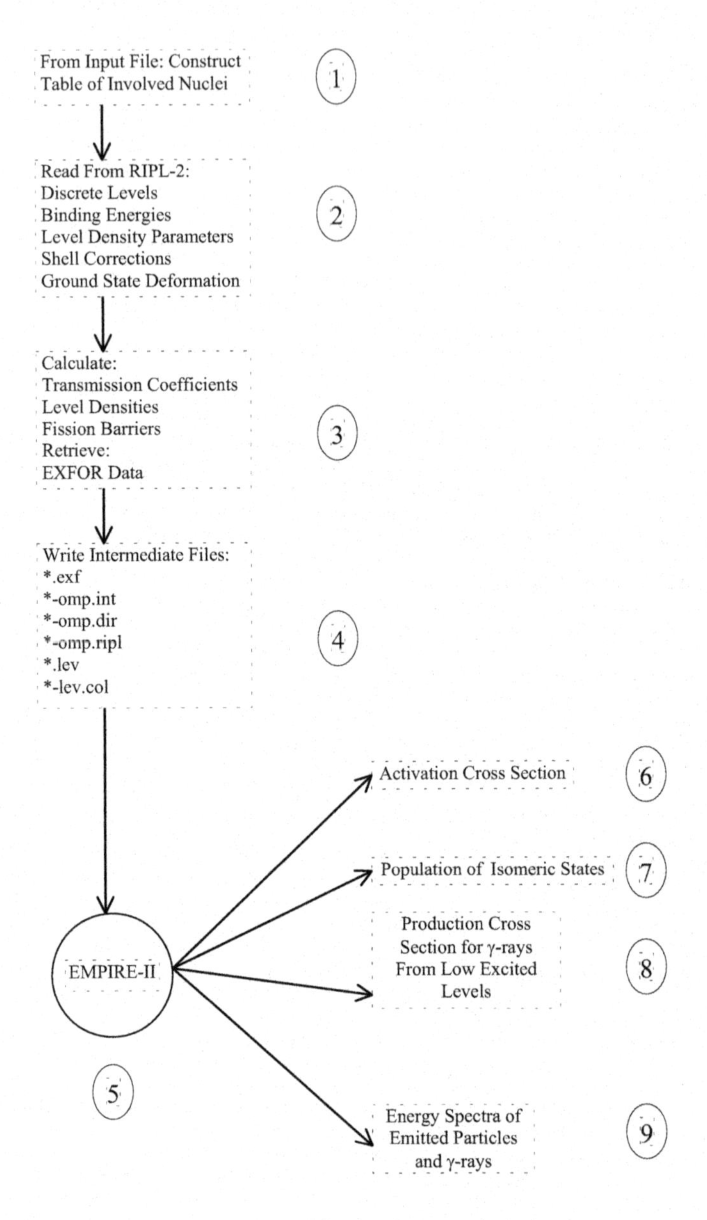

Figure (4-1): Flow Chart of EMPIRE-II Calculations

4-4.11 Input and Output Files

EMPIRE-II has default options for all the input parameter. The mandatory input provided by the user contains the masses and charges of the target and of the projectile and the incident energies. Simply editing the optional input can change the default options and introducing an identifier of the quantity to be changed followed by the label of the desired option.

Among the quantities included in the output file the most important for us through this study are the discrete level populations of the residual nuclei and the discrete γ-transitions. The discrete level populations are calculated after γ-cascade in the continuum but before their de-excitation through discrete transitions. They are directly comparable with the experimental excitation cross-sections measured considering the absolute yield of the gamma rays populating and depopulating the levels.

The γ-cascade between discrete levels starts from the highest discrete level and ends with the first excited state. For each level the population is given, including the complete feeding from the levels above. The γ-lines originating from the level are identified by the energy and spin of the final level and the energy and intensity of the γ-transition. If the branching ratios for the de-population of a given level are not known (are not included in the library LEVELS) the cross-section is trapped at the level and not distributed further [cf. **30**].

4-4.12 Models Compatibility and Priorities

EMPIRE-II allows including different preequilibrium models in a single calculation run, which rises the problem of double counting. In general, ECIS-95 is compatible with DEGAS and HMS, as the latter two do not calculate collective excitations. By the same way ECIS-95 in not compatible with MSD as both include collectivity of discrete levels. However, ECIS-95 and MSD can be combined providing that only the continuum contribution from MSD is retained [cf. **157**].

When combining different models the priorities are the following:

• ECIS-95 provides inelastic scattering to collective levels independently of settings for the remaining models.

• MSD provides inelastic continuum independently of other settings. Inelastic to discrete levels is suppressed if ECIS-95 is active.

• MSC results are taken for inelastic and charge-exchange to continuum if not suppressed by use of DEGAS or HMS.

• DEGAS provides inelastic and charge-exchange to continuum and to discrete levels if MSD and MSC are not active. Otherwise only charge-exchange contribution is used. Gamma emission is used if not provided by MSC.

• HMS provides inelastic and charge-exchange to continuum and to discrete levels if MSD and MSC are not active. Otherwise only charge-exchange contribution is used. Suppresses DEGAS results for particle emission if DEGAS is active, it does not provide γ-rays, thus DEGAS or MSC results are taken.

At incident energies below 20 MeV the best results are expected by combining the four models. At higher incident energies these should be replaced by HMS model, which accounts for the multiple pre-equilibrium emission.

5-1 Calculation of the Theoretical Cross-Sections

5-1.1 Analysis with Code ALICE-91

The calculations were performed using the framework of the geometry dependent hybrid model Blann et al., [cf. **29**,**126**-**131**]. For the pre-equilibrium emission of neutrons and protons in combination with the compound nucleus (CN) calculations were performed using the Weisskopf-Ewing formalism (Weisskopf and Ewing, 1940) [cf. **122**]. While the pre-equilibrium emission component was simulated employing the geometry dependent hybrid (GDH) model Blann, (1972) [cf. **127**] for the subsequent equilibrium emission of neutrons and protons, deuterons and alpha particles. The calculations were based almost on default options in the code Blann, (1988) [cf. **29**]. The initial exciton configuration given by Blann and Vonach, (1983) [cf. **130**] were used for the pre-equilibrium calculations. The initial exciton number, the initial excited neutron number and the initial excited proton number were assumed to be 3.0, 0.8 and 1.2, respectively. The exciton includes the particle and the hole in the nucleus. The initial exciton number (3) is the sum of particles and holes. The initial excited proton number (1.2) corresponds to the initial proton particle number, which includes a portion of the projectile proton. The number 3 - 0.8 - 1.2 = 1 is the initial hole number. The number 0.8 + 1.2 = 2 is the initial particle number. The optical model potential based on the nucleon mean free paths was used to estimate the intranuclear transition rates. Particle binding energies were internally calculated using the Myers and Swiatecki mass formula Myers and Swiatecki, (1967) [cf. **158**]. Inverse cross sections were also given by the code.

The level density parameter "a" is calculated from $a = A/K$, where, A is the mass number of the compound system, and K is a constant, which may be varied to match the excitation functions. The initial exciton numbers are important parameters of pre-equilibrium emission formalism, is defined by $n_o = (p+h)$. It is however, reasonable to assume that an incident proton (particle) in its first interaction excites a particle above the Fermi level leaving behind a hole, i.e., in all two particles and one hole, in creating the initial exciton state. Further, the mean free path (MFP) in these calculations is generated using free nucleon-nucleon scattering cross sections. The calculated MFP for two-body residual interactions may differ from the actual MFP. To account for that, the parameter COST is provided. In this code the MFP is multiplied by (COST+1). As such, by varying the parameter COST, the nuclear MFP can be adjusted to fit the experimental data. The default value of K=10; $n_o = 3(2p+1h)$ and COST=9; for proton induced reactions gave a satisfactory reproduction of the measured data in the earlier analyses of excitation functions Mustafa et al., (1995) [cf. **159**]. The same values of these parameters have been used in the present calculations, and a pairing energy shift of the effective ground state was calculated internally from the Myers and Swiatecki (1967) [cf. **158**] mass formula. Which was introduced together with a shell effect correction shift.

The calculated excitation functions with the above parameters reproduce the measured excitation functions for the ^{nat}Mo(p,x) reactions satisfactorily. However, for the ^{nat}Gd(p,x) reactions the calculations done with these parameters do not give satisfactory agreement with the experimental data. In order to reproduce the experimental data for this reaction the values of K and mean-free-path multiplier are varied and the corresponding excitation functions are in agreement. A value of K=8; with COST=9; has been found to satisfactorily reproduce the excitation function for ^{nat}Gd(p,x) reactions. A larger value of K in case of ^{nat}Gd(p,x) reactions means a relatively smaller value of the level density parameter "a" for the residual nucleus, which is expected in the case of $_{64}$Gd. Further, Weisskopf-Ewing calculations do not include spin effects [cf. **122**].

5-1.2 Analysis with Code EMPIRE-II

Initially, the program calculates the tables of the partial wave transmission coefficients and the spin dependent level densities of nuclei in the decay channels of the composite nucleus. If the (PE) option is selected, the geometry dependents Hybrid model Blann (1972) [cf. **127**]. Subroutine is used to calculate neutron and proton spectra, separately for each partial wave. The (PE) yields are distributed over the continuum of states and the discrete levels of the residual nuclei according to the energy, angular momentum and parity conservation laws.

In a next step the (CN) decay is followed and its contributions are recorded, on local files ascribed to each residual nucleus, and added to the previously calculated (PE) populations.

Multiparticle emission is assumed to take place after the nucleus has reached equilibrium. At each stage, a multiparticle emission process, the decaying nucleus appropriately replaces its ancestor and the decay is treated in terms of the transmission coefficients and level densities. This procedure is repeated until the list of particles emitted in the reaction is exhausted. The population of the discrete levels in the residual nucleus and the spectra of particles and photons emitted at each step of the reaction are printed.

Particles considered in the program are neutrons, protons and alphas, in addition, deuterons in the entrance channel and γ's in the exit channel are included. The (PE) emission does not allow for α–particles. The maximum energy and number of subsequently emitted particles are not restricted in the program. The maximum number of partial waves considered in the calculations is 3D. Only E1, M1, and E2 γ–transitions are allowed to contribute to the γ–cascade. Up to 50 discrete levels for each nucleus taken into account can be used. In the γ–decay of these levels, up to 11 transitions depopulating each level might be concerned.

The input data in the EMPIRE-II code should include the incoming particle energy, the mass and the atomic numbers of the composite nucleus, the mass and the atomic numbers of the incoming particle, the spin of the target nucleus, the number of partial waves accounted in the calculations, the energy step in integration, the number of discrete levels in composite nucleus (5–50), the initial exciton number (in our cases it is set to 3), list of energies followed by list of spins which define discrete levels in composite nucleus, the energy cut off for continuum (taken to be the energy of highest discrete level), the number of discrete levels in the residual nuclei after neutron, proton and α–emission, and the spin cut off factor for n–exciton levels.

The binding energies are taken from Audi and Wapstra (1995) tables [cf. **155**]. The pairing energies and shell corrections are taken from Gilbert and Cameron (1965) [cf. **154**] tables. The inverse cross-sections which represent a parameter of great importance are calculated using the code SCAT2, Bersillon (1991) [cf. **142**]. Several options for the level density parameters are allowed in the EMPIRE-II code. The intranuclear transition rates are calculated from the imaginary optical potential.

The Excitation functions of the reactions ^{nat}Mo and ^{nat}Gd (p,xn) have been performed at proton energies from threshold up to 20 MeV. The calculated cross sections are analyzed in terms of a sum of statistical and pre-equilibrium contributions. Using the EMPIRE-II code-2.19 (Londi) March, (2005) [cf. **30**], which is an energy dependent hybrid model, accounting for the major nuclear reaction mechanisms. Including the Optical Model (OM), the Multi-step Compound, Exciton Model, and the full featured Hauser-Feshbach Model, with a comprehensive parameter library mainly converting nuclear masses, OM data, Discrete nuclear levels, and decay schemes. The Monte Carlo Pre-equilibrium approach has been particularly successful in approximating the experimental values. As the determination of the nuclear level density is of main impact on the results.

The (p,xn) reactions offer a promising approach to the study of the interaction mechanisms of protons beam with target nuclei of Mo, Gd. The excitation functions of this type of reactions exhibit a bell-shaped pattern with a rather broad top. The compound nucleus (CN) picture of the process predicts this behavior but fails to account for the high-energy tail of the excitation functions. This tail is assumed to derive from a non-compound process by means of which the incident particle can eject a neutron from the target nucleus before the (CN) is formed, that is before thermal equilibrium is reached between the nucleons. In this case, the resulting residual nuclei will have low excitation energies and the neutrons emitted in the subsequent evaporation process will be fewer than expected if all the incident energy were to go in (CN) formation. The first problem in the interpretation of the experiments is then to determine the partial cross sections for the two interaction mechanisms. The problems encountered are those of a correct choice of the parameter values and a sensible assessment of the effects of the approximations introduced.

5-1.3 Discrete Level Cross Section

EMPIRE-II is a valuable tool in the spin-parity assignment for discrete levels of residual nuclei populated in nuclear reaction evolving through compound nucleus mechanism. So there is several applications illustrating the method and the code's predictive power.

A major problem in the experimental nuclear spectroscopy is the spin-parity (J^{π}) assignment for the nuclear levels. In direct reactions with hadronic probes, the transferred angular momentum can be deduced from the shape of the angular distributions of the outgoing particles. In gamma-spectroscopy, measurements of gamma-ray angular distributions and conversion electron coefficient are the main tools in establishing a certain range for (J^{π}) values of the nuclear levels.

Another way leading to possible (J^{π}) values for nuclear levels is to exploit the dependence of the compound nucleus decay probabilities on the spin-parity of the discrete levels of the residual nuclei. These decay probabilities enter the level population cross section calculated in the framework of compound nucleus statistical theory of nuclear reactions. By comparing the measured discrete level population cross sections with the calculated one, and assuming different hypothesis for the (J^{π}) of final state a certain restricted range in the (J^{π}) values can be obtained. Several studies of (p,nγ), (α,nγ), (n,n'γ) reactions applied this procedure in selecting certain (J^{π}) values for levels measured by γ-spectroscopical means [cf. **160**-**164**].

In completing this task, a compound nucleus statistical model code has to be used.

There are many model codes, but they usually require an extensive input including particle transmission coefficients provide by direct interaction codes. EMPIRE-II [cf. **30**], is a nuclear reaction code which has not this inconvenience because, it directly accesses the most recent References Input Parameter Library RIPL-2 [cf. **145**], and contains the well-known direct interaction code system, used mainly in nuclear reaction data evaluation.

The goal of this work is to use EMPIRE-II in conjunction with RIPL-2 as a valuable spectroscopic tool for spin-parity assignment to discrete nuclear levels. On the basis of population of the computational instruments, the application of spin-parity assignment is served us in determining the discrete level cross for the predicted nuclear level values. This is one application from several applications that reveal the predictive power of EMPIRE-II.

The EMPIRE-II prediction method of the nuclear level J^{π} Values presented in this work is based on the dependence of the population cross-sections on the spin and parity of the discrete levels of the residual nuclei. The sensitivity of the population cross-section to the variation of the final level spin-parity values depends on the difference between the transmission coefficients entering the Hauser-Feshbach formula when different J^{π} values are considered. It is difficult to make universal assumptions about this sensitivity and the prediction power of this method due to their dependence on several factors such as: reaction type, energy range, odd-even character of the target, spin-parity of the target's ground state, optical potential parameters, etc. [cf. **165**]. Therefore, detailed numerical calculations of the population cross-sections have to be performed case-by-case. EMPIRE-II is an appropriate code system for performing this task especially because it is based on the latest versions of the reaction models, takes into consideration all the open reaction channels, offers many choices for the input parameters and automatically creates the input files. Due to the short execution time of EMPIRE-II, different input parameter values and a large set of J^{π} hypothesis can be tested until the best agreement with the experimental data is obtained. One complete calculation for the reactions presented below on a Pentium IV machine takes only few seconds.

The predictive power of EMPIRE-II was verified in the case of the reaction (p,nγ) especially for discrete γ-ray spectroscopy. The cross section data obtained from ALICE-91 code due to several proton induced reactions on Natural Molybdenum, and Gadolinium targets are shown in the Tables (5-1), (5-3). The Resultant Output Cross Section Values from obtained from EMPIRE-II code due to several proton induced reactions on Natural Molybdenum, and Gadolinium targets are shown in the Tables (5-2), (5-4).

Table (5-1): The Resultant Output Cross Section Values from Several Proton Induced Reactions on Natural Molybdenum Targets, The Calculation of the Total Cross Section from the Individual Contributing Reactions have Performed by Using ALICE-91 Code [cf. **29**].

E_P (MeV)	^{92}Nb σ (mb)	^{95}Tc σ (mb)	^{96}Tc σ (mb)	^{94}Tc σ (mb)	^{99}Tc σ (mb)
1	0.00	0.00	0.00	0.00	0.00
2	0.00	0.00	0.00	0.00	0.00
3	0.00	0.00	0.00	0.00	0.00
4	0.00	1.23	0.00	0.00	0.00
5	0.01	6.69	7.24	0.00	0.00
6	0.04	20.22	22.02	11.10	0.00
7	0.14	41.71	45.20	23.50	0.00
8	0.41	66.23	71.06	37.83	0.00
9	0.77	88.52	94.58	51.06	18.49
10	1.22	106.66	113.76	61.70	43.53
11	1.72	120.99	129.27	70.12	62.50
12	2.28	132.93	142.28	76.96	75.69
13	2.85	139.94	153.62	82.33	85.23
14	3.35	99.02	163.13	87.78	92.26
15	3.61	58.43	170.14	82.79	97.26
16	3.15	34.07	174.97	51.25	100.15
17	2.67	21.01	178.14	32.19	102.08
18	2.25	14.44	180.34	20.44	95.82
19	1.73	11.05	182.71	13.69	79.16
20	1.37	9.49	185.70	9.99	61.25

Table (5-2): The Resultant Output Cross Section Values from Several Proton Induced Reactions on Natural Molybdenum Targets, The Calculation of the Total Cross Section from the Individual Contributing Reactions have Performed by Using EMPIRE-II Code [cf. **30**].

E_P (MeV)	^{92m}Nb σ (mb)	^{94}Tc σ (mb)	^{95m}Tc σ (mb)	^{95g}Tc σ (mb)	$^{96m+g}$Tc σ (mb)	^{99m}Tc σ (mb)
1	0.00	0.00	0.00	0.00	0.00	0.00
2	0.00	0.00	0.00	0.00	0.00	0.00
3	0.00	0.00	0.07	0.08	0.00	0.00
4	0.00	0.00	1.29	1.43	1.14	0.00
5	0.00	0.00	6.44	7.16	7.42	0.00
6	0.00	6.67	14.60	14.75	13.16	0.00
7	0.01	14.10	26.82	29.80	29.52	0.00
8	0.04	22.54	34.60	48.06	50.21	0.16
9	0.09	30.65	39.45	65.00	68.11	14.74
10	0.18	33.40	45.28	76.00	76.86	29.60
11	0.30	37.33	48.56	85.00	84.16	39.61
12	0.44	43.61	53.53	94.00	107.18	49.34
13	0.60	51.00	58.85	107.00	110.87	61.14
14	0.77	68.73	62.49	115.73	106.04	69.41
15	0.93	72.74	64.12	120.14	94.37	65.75
16	1.04	81.37	66.84	123.77	87.57	72.35
17	1.17	88.64	68.96	127.71	85.58	78.30
18	1.38	94.10	71.44	132.30	85.07	82.88
19	1.78	98.91	73.16	135.48	85.23	82.62
20	2.23	102.44	75.55	139.91	85.32	77.12

Table (5-3): The Resultant Output Cross Section Values from Several Proton Induced Reactions on Natural Gadolinium Targets, The Calculation of the Total Cross Section from the Individual Contributing Reactions have Performed by Using ALICE-91 Code [cf. **29**].

E_p (MeV)	**^{152}Tb σ (mb)**	**^{154}Tb σ (mb)**	**^{155}Tb σ (mb)**	**^{156}Tb σ (mb)**	**^{160}Tb σ (mb)**
1	0.00	0.00	0.00	0.00	0.00
2	0.00	0.00	0.00	0.00	0.00
3	0.00	0.00	0.00	0.00	0.00
4	0.00	0.00	0.01	0.01	0.02
5	0.00	0.01	0.15	0.20	0.32
6	0.03	0.13	1.18	1.61	2.27
7	0.02	0.64	4.94	6.84	9.12
8	0.10	1.88	13.66	19.04	24.77
9	0.25	4.03	28.42	39.71	39.74
10	0.50	6.93	48.25	67.55	40.74
11	0.79	10.12	70.00	95.39	33.53
12	1.07	23.66	80.07	105.01	25.59
13	1.30	54.31	67.34	88.23	19.18
14	1.27	85.41	49.28	64.89	14.90
15	1.04	107.44	38.48	45.24	12.25
16	0.79	129.36	26.94	31.11	10.56
17	0.58	145.95	19.54	22.11	9.49
18	0.42	158.98	14.80	16.50	7.58
19	0.31	167.32	11.94	13.14	7.23
20	0.24	169.97	10.18	11.14	7.00

Table (5-4): The Resultant Output Cross Section Values from Several Proton Induced Reactions on Natural Gadolinium Targets, The Calculation of the Total Cross Section from the Individual Contributing Reactions have Performed by Using EMPIRE-II Code [cf. **30**].

E_p (MeV)	$^{152m+g}$Tb σ (mb)	^{154m}Tb σ (mb)	^{154g}Tb σ (mb)	^{155}Tb σ (mb)	^{156}Tb σ (mb)	^{160}Tb σ (mb)
1	0.00	0.00	0.00	0.00	0.00	0.00
2	0.00	0.00	0.00	0.00	0.00	0.00
3	0.00	0.00	0.00	0.00	0.00	0.00
4	0.00	0.00	0.00	0.01	0.00	0.02
5	0.22	0.02	0.03	0.16	0.22	0.26
6	6.01	0.13	0.27	1.10	1.43	1.80
7	25.81	0.59	1.18	4.50	6.04	7.28
8	71.52	1.64	3.28	11.97	16.36	16.76
9	144.06	3.30	6.60	22.22	32.83	16.06
10	172.56	4.06	8.12	29.01	41.63	10.33
11	226.50	5.35	10.69	44.76	52.61	7.31
12	314.12	10.09	20.18	59.91	54.93	5.39
13	339.59	21.46	42.92	66.44	58.38	4.19
14	130.84	34.47	68.93	72.97	57.65	2.09
15	91.76	45.12	90.24	81.85	63.13	1.62
16	46.03	50.25	100.50	83.13	63.87	1.31
17	23.70	49.52	99.04	79.10	59.16	0.86
18	11.77	45.47	90.94	68.99	52.13	0.63
19	7.06	38.77	77.55	60.92	39.65	0.45
20	4.64	33.39	66.79	53.96	31.40	0.37

5-2 The Excitation Function for Produced Radionuclides from Mo Targets

When measuring the cross-sections of the (p,n) reaction on enriched targets, the data do not contain the contributions of the (p,2n) and other reactions with higher ***Q*** values taking place on natural targets. After transforming this data to "natural" composition, the "elemental" cross section data has been derived. That, allowing comparison with other data.

Several reactions contribute to the production of ^{96m}Tc, ^{96g}Tc and ^{96}Nb isotopes as can be seen in Table (2-3). The radioisotope ^{96m}Tc decays to ^{96g}Tc by internal conversion (98%) followed by a low energy and very weak gamma-line not suitable for quantitative assessment. ^{96m}Tc also decays to ^{96}Mo by electron capture and positron emission but the gamma lines emitted during the decay are too weak and/or not independent since ^{96g}Tc and ^{96}Nb also decay to ^{96}Mo producing the same gamma-lines. Separation of the radiation of ^{96m}Tc and ^{96g}Tc is not straightforward. It requires an inconvenient time consuming series of measurement, as detection of the isomeric state through its direct radiation is very difficult (weak and common gamma lines, 2% branching ratio). Detection of the ground state decay is also disturbed by the decay of ^{96}Nb resulting in similar gamma lines. However, determination with good accuracy of the cumulative ^{96mg}Tc cross-section is possible after the complete decay of ^{96}Nb. For monitoring and TLA use measurements have to start after total decay of ^{96}Nb i.e. 3-4 days after EOB (End of Bombardment).

Separation of intense independent gamma lines on the other hand make it possible to determine independently the excitation function of the processes for production of ^{96mg}Tc and ^{96}Nb. Decay analysis shows that the cross-section for production of ^{96m}Tc is significant compared to direct production of ground state, therefore in monitoring it is necessary to wait until the ^{96m}Tc decays completely. The excitation function of the commutative ^{nat}Mo(p,x)^{96mg}Tc process can be measured precisely after complete decay of ^{96}Nb. Three reactions contribute to the production of ^{96}Nb are ^{100}Mo(p,αn), ^{97}Mo(p,2p), and ^{98}Mo(p,2pn) with thresholds 3.82, 9.32, and 18.05 MeV respectively. As ^{96}Nb decays into the ^{96}Mo like ^{96}Tc, independent gamma line is not available to determine the activity of ^{96}Nb separately. The determination of the cross-section for production of ^{96}Nb is however possible by following the decay and decomposing the measured gamma lines. The published measured cross-section is in the range of a few millibarns in the investigated energy region up to 40 MeV [cf. **86**], due to the statistical error and the applied peak de-composition technique the experimental points have an estimated uncertainty of about 30%.

5-2.1 The $^{nat}Mo(p,x)^{92m}Nb$ Reaction

The excitation function of $^{nat}Mo(p,x)^{92m}Nb$ reaction measured in this work is shown in Figure (5-1). The production of ^{92m}Nb from natural Mo targets, total cross section, and thin, thick yields as a function of proton energies, data obtained and their estimated errors are tabulated in Table (5-5). In this work the obtained data are consistent with the available data of N.V. Levkowski (1991) [cf. **82**]. The good agreement, especially in the overlapping energy regions, gives more convenient to our work. ^{92m}Nb, ^{96}Nb radionuclides were identified as reaction product in this work, the production of ^{92m}Nb radionuclide is most likely due to (p,α) reaction type on Mo targets. The production of Nb radioactivities must be considered as a potential source of radionuclidic contamination in the production of Tc radionuclides. Therefore, the radiochemical processing of Mo targets should provide for an efficient separation of Nb radioactivities and or their Zr, and Mo decay products. The comparisons of EMPIRE-II data with that belong to us, show that to some extend seem to be close to our data than ALICE-91 results.

5-2.2 The $^{nat}Mo(p,x)^{94g}Tc$ Reaction

The excitation function of the $^{nat}Mo(p,x)^{94g}Tc$ reaction based on this work is shown in Figure (5-2). The production of ^{94g}Tc from natural Mo targets, total cross section, and thin, thick yields as a function of proton energies, the data obtained and their estimated errors are tabulated in Table (5-6). For comparison the literature data of M. Bonardi et al., (2002); Yu. Zuravlev et al., (1994) [cf. **3**] are also given. Again, the data reported by Bonardi et al., Zuravlev et al., were obtained via measurements on ^{nat}Mo and extrapolation of the results to high isotopic enrichment. The Zuravlev et al., values agree to some extent with our data near the threshold, and at the higher region, the data of Bonardi et al., on the other hand, are in complete agreement with our data. It is shown in the figure too, the agreement between EMPIRE-II and our data and the literature obtained data.

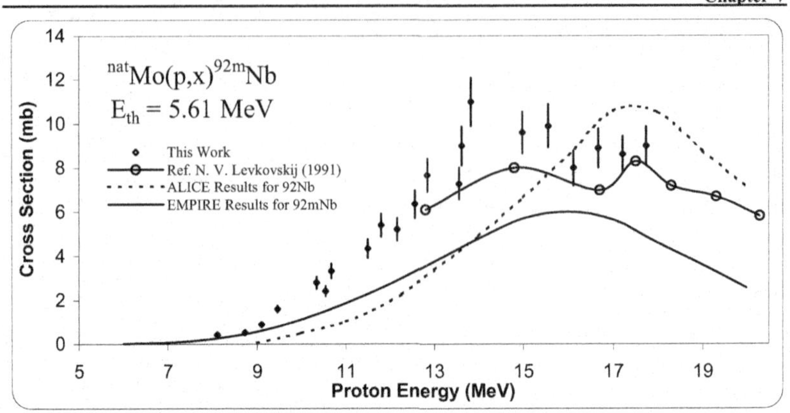

Figure (5-1): The Excitation Function of the $^{nat}Mo(p,x)^{92m}Nb$ Reaction.

Table (5-5). Measured Cross Sections Values for Production of ^{92m}Nb, on Natural Mo Targets at Different Energies.

Incident Energy E_p (MeV)	**$^{nat}Mo(p,x)^{92m}Nb$ σ (mb)**	**Differential Yields mCi/μA·h**	**Integral Yields MBq/C**
17.74±0.75	9.00±0.92	1.94E-04	1.44E+01
17.21±0.73	8.60±0.88	2.00E-04	1.36E+01
16.67±0.71	8.90±0.91	2.03E-04	1.26E+01
16.12±0.68	8.00±0.82	2.06E-04	1.16E+01
15.56±0.66	9.90±1.07	2.06E-04	1.06E+01
14.98±0.64	9.60±1.04	2.03E-04	9.51E+00
13.83±0.59	11.00±1.19	1.89E-04	8.40E+00
13.62±0.58	9.00±0.97	1.84E-04	7.28E+00
13.56±0.57	7.29±0.82	1.75E-04	6.18E+00
12.84±0.54	7.67±0.86	1.54E-04	5.12E+00
12.56±0.53	6.35±0.71	1.34E-04	4.14E+00
12.15±0.52	5.20±0.59	1.11E-04	3.27E+00
11.79±0.50	5.39±0.65	8.88E-05	2.52E+00
11.49±0.49	4.34±0.53	6.88E-05	1.90E+00
10.67±0.45	3.33±0.40	4.94E-05	1.41E+00
10.54±0.45	2.41±0.29	3.67E-05	1.03E+00
10.34±0.44	2.79±0.37	2.65E-05	7.50E-01
9.46±0.40	1.61±0.21	1.75E-05	5.41E-01
9.10±0.39	0.89±0.12	1.22E-05	3.89E-01
8.73±0.37	0.51±0.07	8.40E-06	2.80E-01
8.11±0.34	0.41±0.08	5.67E-06	2.01E-01

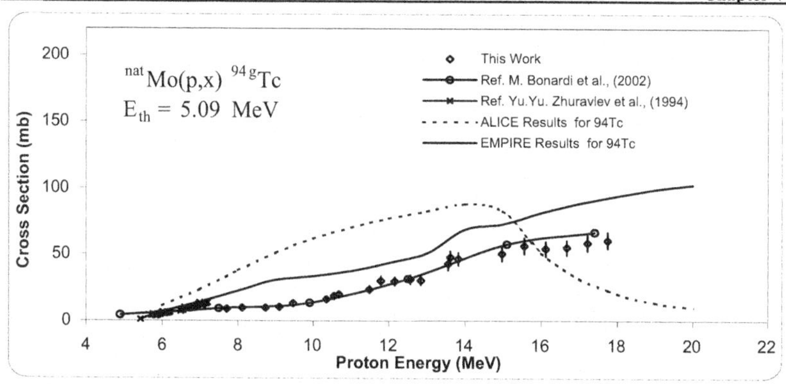

Figure (5-2): The Excitation Function of the $^{nat}Mo(p,x)^{94g}Tc$ Reaction.

Table (5-6). Measured Cross Sections Values for Production of ^{94}Tc, on Natural Mo Targets at Different Energies.

Incident Energy E_p (MeV)	$^{nat}Mo(p,x)^{94}Tc$ σ (mb)	Differential Yields MCi/μA·h	Integral Yields MBq/C
17.74±0.75	60.68±6.22	2.04E+01	4.44E+00
17.21±0.73	58.70±6.02	1.99E+01	4.17E+00
16.67±0.71	55.16±5.66	1.91E+01	3.89E+00
16.12±0.68	54.25±5.56	1.83E+01	3.57E+00
15.56±0.66	56.54±6.11	1.73E+01	3.25E+00
14.98±0.64	50.74±5.49	1.63E+01	2.94E+00
13.83±0.59	46.69±5.05	1.45E+01	2.64E+00
13.62±0.58	47.79±5.17	1.36E+01	2.34E+00
13.56±0.57	42.97±4.84	1.27E+01	2.06E+00
12.84±0.54	30.50±3.43	1.10E+01	1.80E+00
12.56±0.53	31.21±3.51	9.60E+00	1.56E+00
12.15±0.52	29.80±3.35	8.12E+00	1.35E+00
11.79±0.50	30.04±3.64	6.80E+00	1.16E+00
11.49±0.49	23.60±2.86	5.70E+00	1.00E+00
10.67±0.45	19.86±2.41	4.58E+00	8.61E-01
10.54±0.45	18.64±2.26	3.94E+00	7.41E-01
10.34±0.44	16.20±2.16	3.41E+00	6.36E-01
9.46±0.40	13.36±1.78	2.78E+00	5.44E-01
9.10±0.39	10.52±1.40	2.42E+00	4.62E-01
8.73±0.37	9.90±1.32	2.12E+00	3.88E-01
8.11±0.34	9.73±1.99	1.81E+00	3.20E-01
7.71±0.33	8.94±1.83	1.59E+00	2.58E-01

5-2.3 The $^{nat}Mo(p,x)^{95g}Tc$ Reaction

The excitation function of the $^{nat}Mo(p,x)^{95g}Tc$ reaction obtained in this work is shown in Figure (5-3). The production of ^{95g}Tc from natural Mo targets, total cross section, and thin, thick yields as a function of proton energies, the data obtained are tabulated in Table (5-7). The direct formation of the ground, metastable states is assessed in the first hours after EOB by measuring its characteristic γ-line. The ground state has a shorter half-life (^{94g}Tc, $T_{1/2}$=20h) than its metastable state (^{94m}Tc, $T_{1/2}$=61d), this implying that the signal from it with suitable energy is clear quite separate. As no appreciable amount ^{94m}Tc was generated, the resulting cross section data obtained and their estimated errors are entirely determined and represented in Table (5-7). The data of all investigators M. Bonardi et al., (2002); N.V. Levkowski (1991) [cf. **3**,**82**] show on average good agreement over the large energy region, especially with Bonardi et al., but Levkowski data tend to lower data at the tapering end of the excitation function. These data and our measurements are surrounded by an envelope with the higher arm of ALICE-91 and the lower of EMPIRE-II, on average both arms are close to experimental results. Again, the data reported by Bonardi et al., but Levkowski et al., were obtained via measurements on ^{nat}Mo and extrapolation of the results to high isotopic enrichment. EMPIRE-II here uses the default option of the Optical Model Parameters (OMP) which represent the key ingredients in this type of calculations. One of the strongest points of EMPIRE-II code is that it allows a rapid, and easy testing of various sets of OMP. The default option taking into account the following OMP sets from the built in systematics: Wilmore-Hodgson for neutrons, Becchetti-Greenless for protons, and McFadden-Satchler for α-particles [cf. **176**].

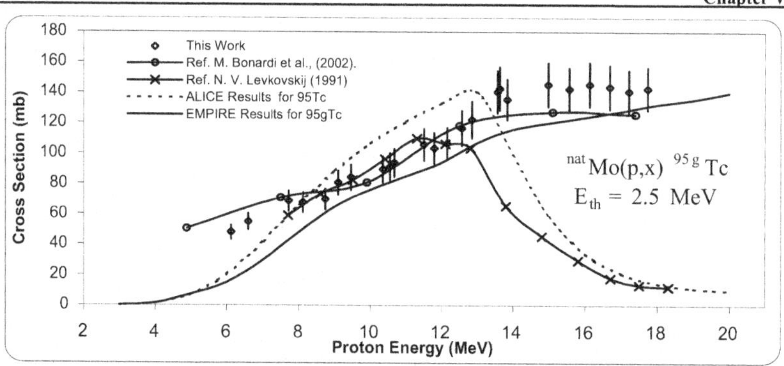

Figure (5-3): The Excitation Function of the $^{nat}Mo(p,x)^{95g}Tc$ Reaction.

Table (5-7). Measured Cross Sections Values for Production of ^{95g}Tc, on Natural Mo Targets at Different Energies.

Incident Energy E_p (MeV)	**$^{nat}Mo(p,x)^{95g}Tc$ σ (mb)**	**Differential Yields MCi/μA·h**	**Integral Yields MBq/C**
17.74±0.75	142.49±14.08	3.00E+00	3.85E+00
17.21±0.73	140.85±13.92	3.02E+00	3.76E+00
16.67±0.71	143.57±14.19	3.00E+00	3.66E+00
16.12±0.68	145.23±14.36	2.96E+00	3.54E+00
15.56±0.66	142.25±14.07	2.91E+00	3.42E+00
14.98±0.64	145.22±14.36	2.83E+00	3.27E+00
13.83±0.59	135.40±14.40	2.61E+00	3.12E+00
13.62±0.58	142.58±14.11	2.54E+00	2.95E+00
13.56±0.57	140.59±13.91	2.46E+00	2.75E+00
12.84±0.54	122.28±12.11	2.24E+00	2.55E+00
12.56±0.53	116.53±11.54	2.09E+00	2.36E+00
12.15±0.52	106.57±10.54	1.92E+00	2.18E+00
11.79±0.50	103.28±10.26	1.76E+00	2.01E+00
11.49±0.49	105.77±1050	1.63E+00	1.85E+00
10.67±0.45	93.82±9.32	1.43E+00	1.70E+00
10.54±0.45	91.27±9.02	1.35E+00	1.55E+00
10.34±0.44	89.52±8.85	1.27E+00	1.41E+00
9.46±0.40	84.32±8.33	1.11E+00	1.28E+00
9.10±0.39	80.78±8.34	1.03E+00	1.15E+00
8.73±0.37	69.84±7.21	9.48E-01	1.03E+00
8.11±0.34	67.67±6.99	8.47E-01	9.11E-01
7.71±0.33	68.59±7.08	7.73E-01	7.97E-01
6.59±0.28	54.97±5.68	6.32E-01	6.89E-01
6.12±0.26	47.96±4.95	5.60E-01	5.84E-01

5-2.4 The $^{nat}Mo(p,x)^{95m}Tc$ Reaction

The production of ^{95m}Tc from natural Mo targets, total cross section, and thin, thick yields as a function of proton energies, the data obtained and their estimated errors are tabulated in Table (5-8). The excitation function of the $^{nat}Mo(p,x)^{95m}Tc$ reaction obtained in this work is shown in Figure (5-4), and compared with the collected literature values. The individual cross sections of reactions constituting $^{nat}Mo(p,x)^{95m}Tc$ product measured on natural target by M. Bonardi et al., (2002); N.V. Levkowski (1991) [cf. 3,82]. We note that the metastable cross section data are rather low than ground state, this character contrary to the usual case. Our measured values and the data obtained from published results on the formation of ^{95m}Tc show good agreement except for the results of Levkowski, that showing tapering tend toward a lower values at higher energies. ALICE-91 could not assign isomeric cross section for a certain level, but EMPIRE-II could able to do that, assigned the higher spin state ^{95g}Tc $(9/2)^+$, Lower spin state ^{95m}Tc $(9/2)^-$. Which, enable us to determine also the isomeric ratio in comparison with experimental isomeric ratio as shown in Figure (5-5). By inspecting Figure (5-4) we see the agreement between EMPIRE-II and our results, of course the taper trend of Levkowski data causes a higher values for the isomeric ratios. Data obtained by M. Bonardi et al., (2002) [cf. 3] show good agreement with our isomeric ratios as shown in Figure (5-5), the Isomeric cross section ratios values for $^{95g}Tc/^{95m}Tc$ are tabulated in Table (5-8). Formations of cross-section ratios are obtained with dividing the higher spin isomer by the low spin isomer.

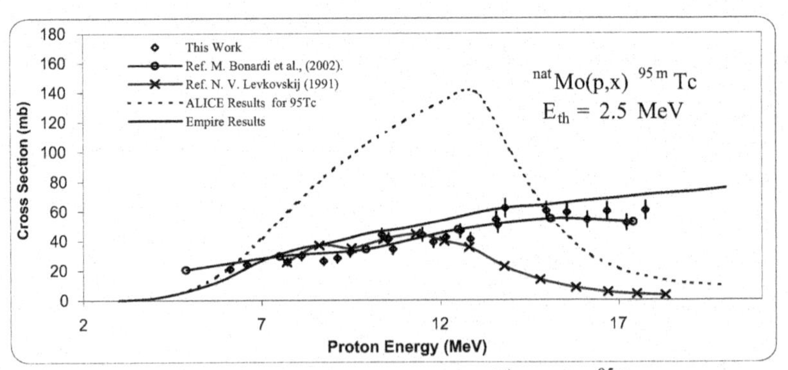

Figure (5-4): The Excitation Function of the $^{nat}Mo(p,x)^{95m}Tc$ Reaction.

Table (5-8). Measured Cross Sections Values for Production of ^{95m}Tc, and Isomeric Ratios on Natural Mo Targets at Different Energies.

Incident Energy E_p (MeV)	$^{nat}Mo(p,x)^{95m}Tc$ σ (mb)	Isomeric Ratios σ^g/σ^m	Differential Yields mCi/μA·h	Integral Yields MBq/C
17.74±0.75	60.58±6.22	2.35	1.76E-02	1.55E+00
17.21±0.73	52.14±5.36	2.70	1.73E-02	1.51E+00
16.67±0.71	59.87±6.15	2.40	1.67E-02	1.48E+00
16.12±0.68	54.27±5.51	2.68	1.61E-02	1.44E+00
15.56±0.66	58.98±5.99	2.41	1.54E-02	1.40E+00
14.98±0.64	60.14±6.11	2.41	1.48E-02	1.35E+00
13.83±0.59	61.99±6.29	2.18	1.35E-02	1.29E+00
13.62±0.58	50.47±5.12	2.83	1.32E-02	1.22E+00
13.56±0.57	54.24±5.50	2.59	1.30E-02	1.14E+00
12.84±0.54	40.97±4.13	2.98	1.21E-02	1.07E+00
12.56±0.53	46.46±4.69	2.51	1.17E-02	9.95E-01
12.15±0.52	42.35±4.27	2.52	1.10E-02	9.23E-01
11.79±0.50	39.19±3.89	2.64	1.05E-02	8.52E-01
11.49±0.49	44.05±4.37	2.40	9.88E-03	7.83E-01
10.67±0.45	34.15±3.39	2.75	8.84E-03	7.15E-01
10.54±0.45	41.23±4.46	2.21	8.34E-03	6.51E-01
10.34±0.44	43.88±4.75	2.04	7.73E-03	5.89E-01
9.46±0.40	32.27±3.49	2.61	6.56E-03	5.30E-01
9.10±0.39	28.29±3.09	2.86	5.76E-03	4.76E-01
8.73±0.37	26.42±2.88	2.64	4.90E-03	4.27E-01
8.11±0.34	29.85±3.26	2.27	4.53E-03	3.83E-01
7.71±0.33	25.98±2.84	2.64	4.23E-03	3.39E-01
6.59±0.28	24.15±2.64	2.28	3.55E-03	2.96E-01
6.12±0.26	21.00±2.29	2.28	3.25E-03	2.54E-01

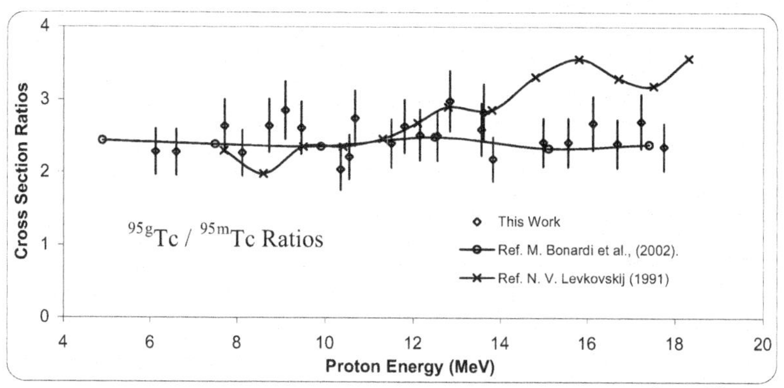

Figure (5-5): The Isomeric Cross Section Ratios Values for $^{95g}Tc/^{95m}Tc$.

5-2.5 The $^{nat}Mo(p,x)^{96m+g}Tc$ Reaction

A literature survey showed that several authors measured cross-sections for the production of the ^{96}Tc isotope using enriched and/or natural molybdenum targets. A summary of the works for the production of ^{96mg}Tc from natural Mo targets, total cross section, and thin, thick yields as a function of proton energies, the data obtained and their estimated errors are tabulated in Table (5-9). The excitation function of the $^{nat}Mo(p,x)^{96m+g}Tc$ reaction obtained in this work is shown in Figure (5-6). Experimental data for more recent works are collected and presented in Figure (5-6). The data reported by S. Takacs et al., (2002) [cf. **86**] present cross-section data for the $^{nat}Mo(p,x)^{96mg}Tc$ process in their paper measured the stacked samples were irradiated in vacuum at an external beam line of a cyclotron accelerator with high energy precision. Data are given up to 38 MeV and are in fair agreement with our data, in the match energy range. M. Bonardi et al., [cf. **3**] presented numerical data measured on natural molybdenum targets up to 43.7 MeV. The type of cross-section they present concerns cumulative elemental cross-section for the $^{nat}Mo(p,x)^{96mg}Tc$ process. The trend of the high-energy tail of the excitation function is reproduced by the inclusion of the pre-equilibrium calculation of ALICE-91 code while the cross section obtained by EMPIRE-II disagree with the experimental data, this disagreement reveal that the good selection of optical model parameters produces particles transmission coefficient which is responsible elevating the cross sections results. As well as the level densities calculation could uses different models or even a mix such as Fermi Gas model (FG) + Hartree-Fock approach (BCS) which is the default option for level density functions in EMPIRE-II, which consider the deformation-dependent collective effects. For the nuclei involved in the reaction induced by protons with a few MeV show more suitable with Gilbert-Cameron density function; its parameters are adjusted to the cumulative number of low-lying states and to the mean level distance at neutron binding energy. In discrete-level spectroscopy with light particle induced reactions in general and particularly not in this case ^{96g}Tc (7^+), ^{96m}Tc (4^+), the residual nuclei are not populated in states with very high excitation energies. Therefore, the level density functions describing the continuum spectra are not crucial for the calculated cross sections. The total cross section ($\sigma_m+\sigma_g$) of the reaction producing ^{96}Tc which is a better agreement in the high energy tail region of the excitation function is achieved with ALICE-91, using all default parameters with a 2p-1h initial exciton configuration of the Hybrid model (Blann, 1971) [cf. **126**], used Superfluid nuclear model for level density function.

Table (5-9). Measured Cross Sections Values for Production of $^{96m+g}Tc$, on Natural Mo Targets at Different Energies.

Incident Energy E_p (MeV)	$^{nat}Mo(p,x)^{96m+g}Tc$ σ (mb)	Differential Yields MCi/μA·h	Integral Yields MBq/C
17.74±0.75	117.00±11.70	8.95E-02	7.47E-01
17.21±0.73	115.14±11.51	9.23E-02	7.26E-01
16.67±0.71	120.00±12.00	9.41E-02	7.02E-01
16.12±0.68	111.25±11.13	9.57E-02	6.73E-01
15.56±0.66	121.34±12.13	9.70E-02	6.42E-01
14.98±0.64	125.74±12.57	9.78E-02	6.08E-01
13.83±0.59	154.72±15.47	9.43E-02	5.72E-01
13.62±0.58	148.95±14.90	9.63E-02	5.34E-01
13.56±0.57	134.97±13.50	9.89E-02	4.95E-01
12.84±0.54	151.37±15.14	9.57E-02	4.55E-01
12.56±0.53	154.47±15.45	9.47E-02	4.14E-01
12.15±0.52	144.21±14.42	9.15E-02	3.73E-01
11.79±0.50	134.29±13.43	8.75E-02	3.31E-01
11.49±0.49	128.42±12.84	8.25E-02	2.91E-01
10.67±0.45	111.67±11.17	7.28E-02	2.51E-01
10.54±0.45	109.15±10.92	6.71E-02	2.13E-01
10.34±0.44	117.59±11.76	6.01E-02	1.78E-01
9.46±0.40	108.23±10.82	4.92E-02	1.46E-01
9.10±0.39	92.18±9.22	4.16E-02	1.18E-01
8.73±0.37	78.52±7.85	3.44E-02	9.27E-02
8.11±0.34	67.99±6.80	2.69E-02	7.10E-02
7.71±0.33	43.80±4.38	2.11E-02	5.28E-02

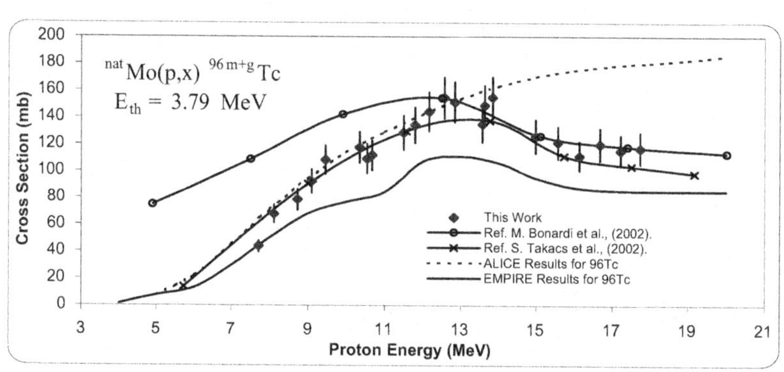

Figure (5-6): The Excitation Function of the $^{nat}Mo(p,x)^{96m+g}Tc$ Reaction.

5-2.7 The $^{nat}Mo(p,x)^{99m}Tc$ Reaction

Three reactions contribute to the production of $^{nat}Mo(p,x)^{99m}Tc$ by direct way are $^{98}Mo(p,\gamma)$, $^{100}Mo(p,2n)$, and indirect way by $^{100}Mo(p,pn)$. The excitation function of the $^{nat}Mo(p,x)^{99m}Tc$ reaction obtained in this work is shown in Figure (5-8). The production of ^{96mg}Tc from natural Mo targets, total cross section, and thin, thick yields as a function of proton energies, the data obtained and their estimated errors are tabulated in Table (5-10). The excitation function of the $^{98}Mo(p,\gamma)^{99m}Tc$ reaction cross section is very small and the errors are large as reported by B. Scholten et al., (1999) [cf. **44**]. This reaction is of no practical relevance for the production of ^{99m}Tc. This result obtained by Scholten is contrary to that previously claimed by Lagunas-Solar et al. (1991) [cf. **49**] who supposed a high (p,γ) cross section, is responsible for ^{99m}Tc production. Presumably, "as Scholten proposed due to some impurity effect. He also noted that, capture reactions in the MeV region have a cross section typically of the order of 1 mb, both in (n,γ) and (p,γ) processes [cf. **44**].

Possibly the whole trend may be a reaction of the $^{100}Mo(p,2n)^{99m}Tc$ reaction (on the 9.63% ^{100}Mo present in the naturally highly pure Mo sample). In contrast, our results on the $^{nat}Mo(p,x)^{99m}Tc$ reaction are higher than the data reported by Scholten et al., (1999) [cf. **44**] in the same time are in good agreement with others Lagunas-Solar et al., (1991); Levkowski et al., (1991) [cf. **49**,**82**] in the same energy region. It is shown from the figure (5-8) that, the cross sections produced by ALICE-91 code is away from the whole experimental date. However, it was for our data, and the published literature. This is because ALICE-91 could not able to predict a cross section for isomeric state. But only for the whole ^{99}Tc produced. EMPIRE-II is capable to predict, assign the level spin and parity for isomeric state responsible mainly for this process and evaluate the cross section pertained to it.

Lagunas-Solar et al., (1991) [cf. **49**] postulated that the production of ^{99m}Tc could be due to two distinct reaction channels. The excitation function shows the first maximum at 19 MeV and was ascribed to the reaction channel $^{98}Mo(p,\gamma)^{99m}Tc$ (Q=-6.36 MeV). It was strongly argued by Lagunas-Solar (1993) that the $^{98}Mo(p,\gamma)^{99m}Tc$ nuclear reaction with 20→10 MeV protons on enriched ^{98}Mo was viable and would have the added advantage of potentially reducing costs of cyclotron installation and operation. The second maximum in the cross section appeared at 42 MeV and was suggested to be due to the $^{100}Mo(p,2n)^{99m}Tc$ (Q=-7.85 MeV) reaction. Scholten measurements using highly enriched Mo as target material depict clearly that the peak at $E_p \approx 17$ MeV is due to the $^{100}Mo(p,2n)^{99m}Tc$ reaction and not the $^{98}Mo(p,\gamma)$-process at higher energies no other peak was observed. The second maximum in the excitation function at higher energies reported by Lagunas-Solar et al. (1991) [cf. **49**] is thus obscure.

It should be emphasized that the $^{100}Mo(p,2n)^{99m}Te$ reaction cannot be compared to a normal (p,2n) reaction in this mass region, since the produced activity is an isomeric state. The systematics is generally valid for total (p,xn) cross sections but not for the formation of higher spins isomers. Even detailed statistical model calculations, incorporating precompound model and nuclear structure effects are often incapable of reproducing the isomeric cross section Nagame et al., (1994) [cf. **24**]. An accurate experimental database is thus crucial to consider the feasibility of the $^{100}Mo(p,2n)^{99m}Te$ reaction for a possible production of ^{99m}Tc at a cyclotron. A limiting factor in this regard would be the level of co-produced long-lived ^{99g}Tc impurity. Experimentally this is very hard to determine and was outside the scope of the present work.

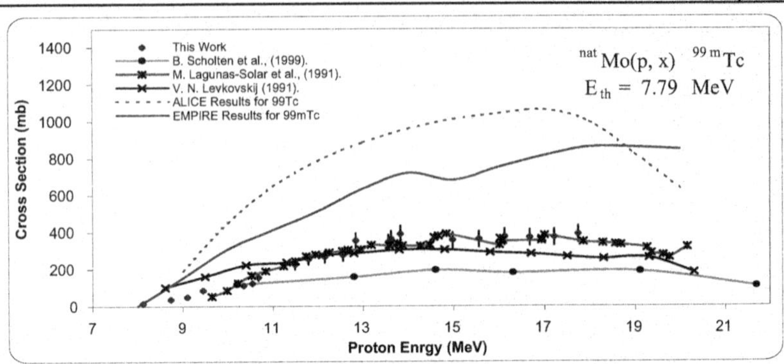

Figure (5-7): The Excitation Function of the $^{nat}Mo(p,x)^{99m}Tc$ Reaction.

Table (5-10). Measured Cross Sections Values for Production of ^{99m}Tc on Natural Mo Targets at Different Energies.

Incident Energy E_p (MeV)	**$^{nat}Mo(p,x)^{99m}Tc$ σ (mb)**	**Differential Yields mCi/μA·h**	**Integral Yields MBq/C**
17.74±0.75	388.00±39.80	8.24E+00	2.48E+01
17.21±0.73	372.16±38.17	8.14E+00	2.31E+01
16.67±0.71	369.42±37.89	7.96E+00	2.14E+01
16.12±0.68	374.00±38.36	7.77E+00	1.98E+01
15.56±0.66	361.68±39.10	7.57E+00	1.81E+01
14.98±0.64	355.44±38.43	7.36E+00	1.63E+01
13.83±0.59	388.35±41.98	6.85E+00	1.46E+01
13.62±0.58	362.75±39.22	6.78E+00	1.29E+01
13.56±0.57	344.47±38.78	6.71E+00	1.11E+01
12.84±0.54	352.39±39.67	6.17E+00	9.36E+00
12.56±0.53	275.45±31.01	5.62E+00	7.67E+00
12.15±0.52	265.52±29.89	4.81E+00	6.09E+00
11.79±0.50	254.27±30.85	3.95E+00	4.69E+00
11.49±0.49	224.27±27.21	3.16E+00	3.51E+00
10.67±0.45	156.31±18.96	2.36E+00	2.54E+00
10.54±0.45	123.24±14.95	1.82E+00	1.76E+00
10.34±0.44	111.65±14.88	1.35E+00	1.15E+00
9.46±0.40	85.51±11.40	8.85E-01	6.89E-01
9.10±0.39	50.12±6.68	5.60E-01	3.59E-01
8.73±0.37	36.52±4.87	2.87E-01	1.42E-01
8.11±0.34	14.72±3.01	5.87E-02	2.55E-02

5-3 The Excitation Functions for Produced Radionuclides from Gd Targets

It has been found by surveying for all the available literature data, that no experimental data is existing for this reaction type on natural or enriched Gd targets in our energy range. So, our data represent the first one. We will then compare our results with those obtained by theoretical code evaluation using ALICE-91 and EMPIRE-II codes. This context mention above will applying to all reactions induced by proton on natural Gadolinium targets, except for the reaction ^{nat}Gd(p,x)^{160}Tb. There is found one available data carried out by C. Birattari et al., (1973) [cf. **78**].

5-3.1 The ^{nat}Gd(p,x)$^{152m+g}$Tb Reaction

The excitation function of the ^{nat}Gd(p,x)^{152g}Tb reaction obtained in this work is shown in Figure (5-8). The production of ^{152}Tb from natural Gd targets, total cross section, and thin, thick yields as a function of proton energies, data obtained and their estimated errors are tabulated in Table (5-11). However, in the energy scale of the ^{nat}Gd(p,x)^{152g}Tb the cross sections obtained for the entire ^{152}Tb production below 20 MeV responsible from ^{152}Gd(p,n)^{152g}Tb only, has $\boldsymbol{Q}$ value of –4.77 MeV. The earlier contributing cross sections (measured on natural targets) are also presented and for comparison they are summarized and normalized to isotopic enrichment to get "isotopic" cross sections for comparison with code calculation. Our measured isotopic production cross sections and the data obtained from the separate individual reaction of EMPIRE-II results, after conversion is in close agreement. While the ALICE-91 output data are in agreement with our measurements at low proton energies below 9 MeV, but in the rest of energy region its data are elevated in disagreement manner.

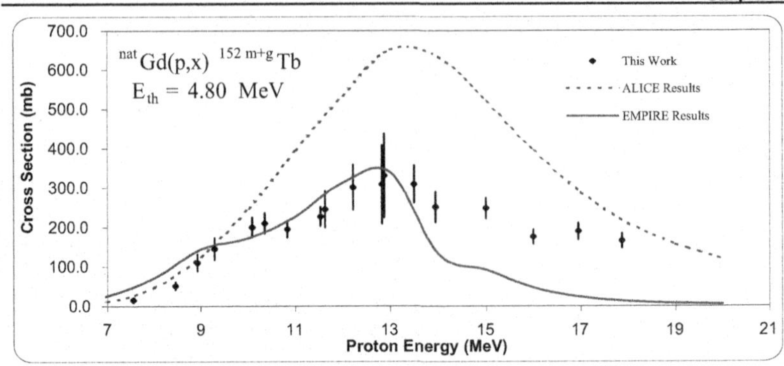

Figure (5-8): The Excitation Function of the $^{nat}Gd(p, x)^{152m+g}Tb$ Reaction.

Table (5-11). Measured Cross Sections Values for Production of $^{152m+g}Tb$, on Natural Gd Targets at Different Energies.

Incident Energy E_p (MeV)	$^{nat}Gd(p,x)^{152m+g}Tb$ σ (mb)	Differential Yields mCi/μA·h	Integral Yields MBq/C
17.87±0.76	165.59±19.27	2.466E-02	1.68E+01
16.95±0.72	189.25±22.02	2.442E-02	1.62E+01
15.99±0.68	175.56±18.34	2.389E-02	1.55E+01
14.99±0.64	248.27±25.94	2.343E-02	1.48E+01
13.94±0.59	251.42±39.18	2.306E-02	1.41E+01
13.50±0.57	309.77±48.27	2.396E-02	1.33E+01
12.87±0.55	331.70±107.38	2.489E-02	1.24E+01
12.82±0.54	309.15±100.08	2.736E-02	1.15E+01
12.21±0.52	302.04±57.96	2.884E-02	1.05E+01
11.63±0.49	245.56±47.12	2.981E-02	9.36E+00
11.53±0.49	227.03±24.78	3.050E-02	8.14E+00
10.82±0.46	195.12±21.29	2.745E-02	6.88E+00
10.34±0.44	210.12±26.94	2.355E-02	5.67E+00
10.08±0.43	200.16±25.67	1.986E-02	4.59E+00
9.29±0.39	145.20±28.11	1.569E-02	3.65E+00
8.92±0.38	110.28±21.35	1.296E-02	2.84E+00
8.45±0.36	51.32±9.93	1.058E-02	2.15E+00
7.55±0.32	15.56±3.01	8.073E-03	1.55E+00

5-3.2 The $^{nat}Gd(p,x)^{154g}Tb$ Reaction

The excitation function of the $^{nat}Gd(p,x)^{154g}Tb$ reaction obtained in this work is shown in Figure (5-9). The resulting cross section data in comparison with the currently obtained from ALICE-91 and EMPIRE-II codes are presented in Figure (5-8). The production of ^{154g}Tb from natural Gd targets, total cross section, and thin, thick yields as a function of proton energies, data obtained and their estimated errors are tabulated in Table (5-12). Three reactions contribute to the formation of ^{154}Tb as mentioned above, the $^{154}Gd(p,n)^{154}Tb$, $^{155}Gd(p,2n)^{154}Tb$, and $^{156}Gd(p,3n)$ ^{154}Tb, having ***Q***-values of –4.34, –10.78, and –19.32 MeV, respectivily. The cross-sections derived from the two irradiation are in good agreement in the overlapping energy region and they are also fit well to the resulting values given by the ALICE-91, EMPIRE-II codes. The experimental cross section data are in good agreemen with EMPIRE-II calculationst, there are a small shift towards higher values could be observed between the excitation function reproduced by ALICE-91, and EMPIRE-II codes. This shift because the data obtained from ALICE-91 represent the total cross section that produces the radionuclide ^{154}Tb, but EMPIRE-II gives the level excitation energy responsible for the ^{154g}Tb in ground state, with spin $J^{\pi}(0)$.

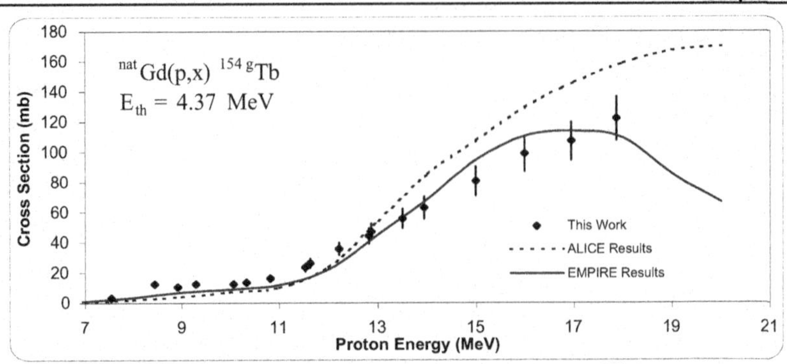

Figure (5-9): The Excitation Function of the $^{nat}Gd(p, x)^{154g}Tb$ Reaction.

Table (5-12). Measured Cross Sections Values for Production of ^{154g}Tb, on Natural Gd Targets at Different Energies.

Incident Energy E_p (MeV)	$^{nat}Gd(p,x)^{154g}Tb$ σ (mb)	Differential Yields mCi/μA·h	Integral Yields MBq/C
17.87±0.76	122.0±16.01	5.27E+00	3.30E+00
16.95±0.72	107.0±14.07	4.72E+00	2.96E+00
15.99±0.68	99.00±11.81	4.27E+00	2.64E+00
14.99±0.64	80.85±9.65	3.81E+00	2.33E+00
13.94±0.59	63.11±7.45	3.31E+00	2.04E+00
13.50±0.57	55.77±6.58	2.96E+00	1.76E+00
12.87±0.55	47.15±6.29	2.58E+00	1.51E+00
12.82±0.54	44.25±5.91	2.34E+00	1.28E+00
12.21±0.52	35.78±4.56	2.00E+00	1.06E+00
11.63±0.49	26.25±3.34	1.68E+00	8.74E-01
11.53±0.49	23.46±2.31	1.42E+00	7.06E-01
10.82±0.46	16.24±1.60	1.09E+00	5.63E-01
10.34±0.44	13.50±1.61	8.04E-01	4.47E-01
10.08±0.43	12.32±1.47	5.82E-01	3.57E-01
9.29±0.39	12.51±1.81	3.98E-01	2.90E-01
8.92±0.38	10.62±1.54	3.08E-01	2.40E-01
8.45±0.36	12.31±1.78	2.76E-01	2.00E-01
7.55±0.32	3.31±0.48	2.58E-01	1.62E-01

5-3.3 The $^{nat}Gd(p,x)^{154m}Tb$ Reaction

The excitation function of the $^{nat}Gd(p,x)^{154m}Tb$ reaction obtained in this work is shown in Figure (5-10). The production of ^{154m}Tb from natural Gd targets, total cross section, and thin, thick yields as a function of proton energies, data obtained and their estimated errors are tabulated in Table (5-13). The experimental data are compared with the output of EMPIRE-II code result. The individual cross sections of the $^{154}Gd(p,n)^{154}Tb$, $^{155}Gd(p,2n)^{154}Tb$, and $^{156}Gd(p,3n)^{154}Tb$ carried out on natural target, having ***Q***-values of –4.34, –10.78, –19.32 MeV, respectivily. Our measured cross section data on the formation of ^{154m}Tb are in good agreement with the predicted cross sections from EMPIRE-II, but ALICE-91 code failed to predict the reaction cross sections, that because ALICE-91 can not determine the cross section for isomeric level. EMPIRE-II gives the level excitation energy responsible for the ^{154m}Tb in two metastable states. The lower spin state of the first metastable state with spin $J^{\pi}(3^{-})$, half-life 9.4 h, and the ground state with higher spin $J^{\pi}(7^{-})$, half-life 22.7 h. The cross sections for the first metastable state with spin $J^{\pi}(3^{-})$ and half life 9.4 h have been calculated. EMPIRE-II could enable us to determine the isomeric ratio in comparison with experimental isomeric ratio as shown in Figure (5-11). The Isomeric cross section ratio values for $^{154g}Tb/^{154m}Tb$ are tabulated in Table (5-12). Formations of cross-section ratios are obtained with dividing the higher spin isomer by the low spin isomer.

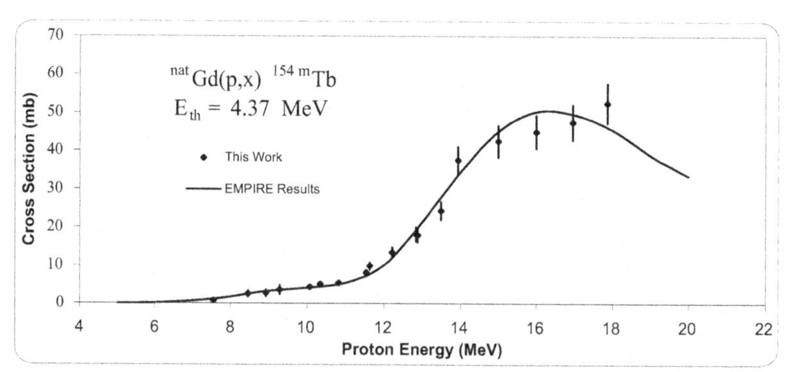

Figure (5-10): The Excitation Function of the $^{nat}Gd(p, x)^{154m}Tb$ Reaction.

Table (5-13). Measured Cross Sections Values for Production of ^{154m}Tc, and Isomeric Ratios on Natural Gd Targets at Different Energies.

Incident Energy E_p (MeV)	^{nat}Gd(p,x)^{154m}Tb σ (mb)	Isomeric Ratios σ^g/σ^m	Differential Yields mCi/μA·h	Integral Yields MBq/C
17.87±0.76	52.50±5.19	2.33	1.34E+01	3.64E+00
16.95±0.72	47.50±4.70	2.26	1.28E+01	3.26E+00
15.99±0.68	45.00±4.46	2.20	1.19E+01	2.89E+00
14.99±0.64	42.50±4.21	1.90	1.09E+01	2.52E+00
13.94±0.59	37.50±3.73	1.68	9.89E+00	2.16E+00
13.50±0.57	24.27±2.41	2.30	9.21E+00	1.80E+00
12.87±0.55	17.97±1.90	2.62	8.27E+00	1.46E+00
12.82±0.54	18.20±1.92	2.43	7.44E+00	1.14E+00
12.21±0.52	13.43±1.46	2.66	5.91E+00	8.53E-01
11.63±0.49	9.84±1.07	2.67	4.18E+00	6.11E-01
11.53±0.49	8.04±0.80	2.92	2.84E+00	4.32E-01
10.82±0.46	5.36±0.53	3.03	1.83E+00	3.08E-01
10.34±0.44	5.08±0.51	2.66	1.24E+00	2.24E-01
10.08±0.43	4.35±0.43	2.83	8.79E-01	1.65E-01
9.29±0.39	3.61±1.25	3.47	6.04E-01	1.21E-01
8.92±0.38	2.71±0.94	3.92	4.38E-01	8.85E-02
8.45±0.36	2.54±0.88	4.84	3.13E-01	6.40E-02
7.55±0.32	0.72±0.25	4.59	2.09E-01	4.55E-02

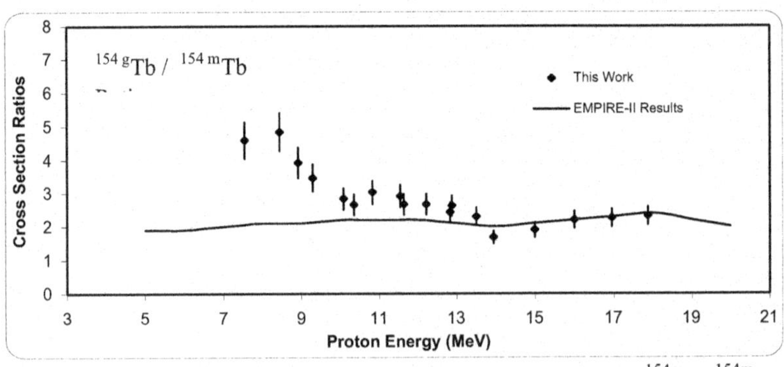

Figure (5-11): The Isomeric Cross Section Ratios Values for ^{154g}Tb/^{154m}Tb.

5-3.4 The $^{nat}Gd(p,x)^{155}Tb$ Reaction

The excitation function of the $^{nat}Gd(p,x)^{155}Tb$ reaction obtained in this work is shown in Figure (5-12). The total cross section for production of ^{155}Tb from natural Gd targets, thin and thick yields as a function of proton energies, data obtained, and their estimated errors are tabulated in Table (5-14). Three reactions contribute to the formation of ^{155}Tb by proton bombardment below 20 MeV on a natural Gd target. There are three nuclear reactions could contribute in the production of ^{155}Tb, the ***Q***-values of the reactions $^{155}Gd(p,n)^{155}Tb$, $^{156}Gd(p,2n)^{155}Tb$, and $^{157}Gd(p,3n)^{155}Tb$ are –1.60, –10.14 and –16.50 MeV, respectively. The nuclear data (half-life, gamma energy and intensity) were taken from [cf. **4**]. The target foils were measured after EOB in a well-calibrated geometry of a HPGe detector. The two sets of experimental points show a good agreement in the overlapping energy region and are also in good agreement with the data resulting from ALICE-91 and EMPIRE-II codes after application of the above discussed energy scale and beam current corrections. The measured experimental points are shown in Figure (5-12) together with the excitation function reproduced from the ALICE-91 and EMPIRE-II codes.

5-3.5 The $^{nat}Gd(p,x)^{156}Tb$ Reaction

The excitation function of the $^{nat}Gd(p,x)^{156}Tb$ reaction obtained in this work is shown in Figure (5-13). Numerical date of the production of ^{156}Tb from natural Gd targets, total cross section, and thin, thick yields as a function of proton energies, data obtained and their estimated errors are tabulated in Table (5-15). In the production of ^{156}Tb only the three proton induced reactions $^{156}Gd(p,n)^{156}Tb$, $^{157}Gd(p,2n)^{156}Tb$, and $^{158}Gd(p,3n)^{156}Tb$ are involved, with ***Q***-values of –3.23, –9.59 and –17.52 MeV, respectively. After evaluating the experimental data of the two irradiated stacks good agreement was found in the overlapping energy region for the measured cross sections. The measured experimental points are shown in Figure (5-13) together with the excitation function reproduced from the codes calculations by ALICE-91 and EMPIRE-II.

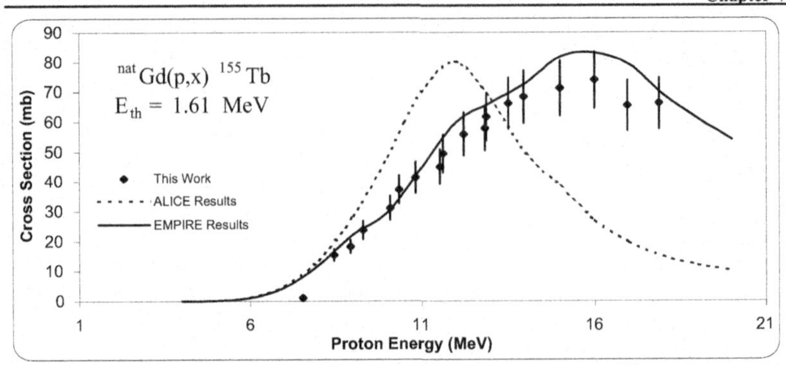

Figure (5-12): The Excitation Function of the $^{nat}Gd(p, x)^{155}Tb$ Reaction.

Table (5-14). Measured Cross Sections Values for Production of ^{155}Tc, on Natural Gd Targets at Different Energies.

Incident Energy E_p (MeV)	$^{nat}Gd(p,x)^{155}Tb$ σ (mb)	Differential Yields mCi/μA·h	Integral Yields MBq/C
17.87±0.76	66.24±7.62	9.101E-02	5.78E-01
16.95±0.72	65.45±7.53	9.029E-02	5.45E-01
15.99±0.68	74.00±8.40	8.753E-02	5.10E-01
14.99±0.64	71.25±8.09	8.382E-02	4.75E-01
13.94±0.59	68.42±7.74	7.902E-02	4.38E-01
13.50±0.57	66.22±7.49	7.683E-02	4.01E-01
12.87±0.55	61.80±6.83	7.266E-02	3.64E-01
12.82±0.54	57.90±6.40	7.081E-02	3.26E-01
12.21±0.52	55.89±6.39	6.498E-02	2.93E-01
11.63±0.49	49.45±5.65	5.868E-02	2.57E-01
11.53±0.49	45.05±4.81	5.435E-02	2.22E-01
10.82±0.46	41.36±4.42	4.695E-02	1.90E-01
10.34±0.44	37.35±4.04	4.074E-02	1.60E-01
10.08±0.43	31.13±3.36	3.560E-02	1.33E-01
9.29±0.39	23.77±6.17	2.903E-02	1.09E-01
8.92±0.38	18.25±4.74	2.424E-02	8.79E-02
8.45±0.36	15.45±4.01	1.947E-02	6.93E-02
7.55±0.32	1.05±0.27	1.415E-02	5.36E-02

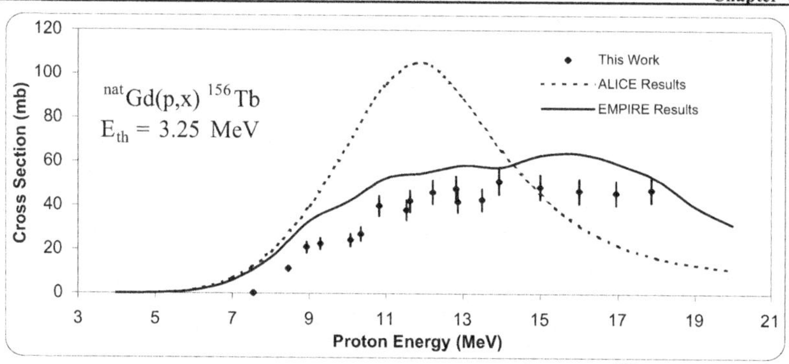

Figure (5-13): The Excitation Function of the $^{nat}Gd(p,x)^{156}Tb$ Reaction.

Table (5-15). Measured Cross Sections Values for Production of ^{156}Tb, on Natural Gd Targets at Different Energies.

Incident Energy E_p (MeV)	**$^{nat}Gd(p,x)^{156}Tb$ σ (mb)**	**Differential Yields MCi/μA·h**	**Integral Yields MBq/C**
17.87±0.76	47.12±4.69	6.439E-02	4.24E-01
16.95±0.72	45.85±4.56	6.165E-02	4.01E-01
15.99±0.68	46.71±4.93	5.806E-02	3.77E-01
14.99±0.64	48.51±5.12	5.455E-02	3.53E-01
13.94±0.59	51.22±5.48	5.104E-02	3.29E-01
13.50±0.57	42.73±4.57	4.985E-02	3.04E-01
12.87±0.55	41.77±4.26	4.789E-02	2.79E-01
12.82±0.54	48.00±4.90	4.790E-02	2.53E-01
12.21±0.52	46.28±4.75	4.547E-02	2.28E-01
11.63±0.49	42.11±4.32	4.270E-02	2.02E-01
11.53±0.49	37.97±3.75	4.119E-02	1.77E-01
10.82±0.46	39.93±3.95	3.700E-02	1.52E-01
10.34±0.44	27.00±2.68	3.320E-02	1.29E-01
10.08±0.43	24.27±2.41	2.977E-02	1.07E-01
9.29±0.39	22.66±6.51	2.468E-02	8.66E-02
8.92±0.38	21.00±6.03	2.080E-02	6.83E-02
8.45±0.36	11.37±3.26	1.683E-02	5.23E-02
7.55±0.32	0.11±0.03	1.242E-02	3.86E-02

5-3.6 The $^{160}Gd(p,n)^{160}Tb$ Reaction

The excitation function of the $^{160}Gd(p,n)^{160}Tb$ reaction obtained in this work is shown in Figure (5-14). The obtained experimental data in this work of the total cross section for the production of ^{160}Tb from natural Gd targets and thin, thick yields as a function of proton energies, data obtained and their estimated errors are tabulated in Table (5-16). The ^{160}Tb isotope is produced by the $^{160}Gd(p,n)^{160}Tb$ reaction below 20 MeV. The ***Q***-value of the reactions is –0.888 MeV. Our experimental data, and the unique available experimental data from the literature of C. Birattari et al., (1973) [cf. **78**], are in good agreement in the overlapping energy region and also are in good correlation with the resulting values given by the EMPIRE-II. On the other hand, after applying the small corrections on the beam current and the energy scale we found some disagreement between the ALICE-91 reproduced data and our experimental data, especially at the higher energy region. The experimental data are in good agreement in the low energy below ~8 MeV. The measured experimental points are shown in Figure (5-14) together with the excitation function reproduced from the ALICE-91 and EMPIRE-II codes.

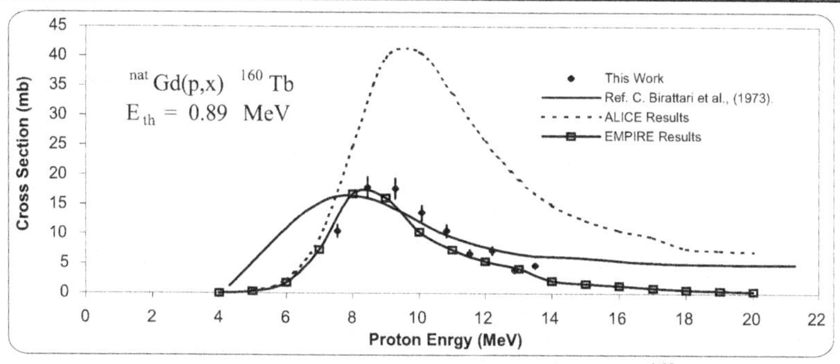

Figure (5-14): The Excitation Function of the $^{nat}Gd(p,x)^{160}Tb$ Reaction.

Table (5-16). Measured Cross Sections Values for Production of ^{160}Tb, on Natural Gd Targets at Different Energies.

Incident Energy E_p (MeV)	**$^{160}Gd(p,n)^{160}Tb$ σ (mb)**	**Differential Yields mCi/μA·h**	**Integral Yields MBq/C**
17.87±0.76	18.00±5.84	2.879E-02	3.49E+01
16.95±0.72	17.00±5.51	2.796E-02	3.48E+01
15.99±0.68	17.50±6.57	2.655E-02	3.45E+01
14.99±0.64	19.00±7.13	2.510E-02	3.41E+01
13.94±0.59	19.18±4.45	2.365E-02	3.36E+01
13.50±0.57	21.42±4.97	2.340E-02	3.31E+01
12.87±0.55	17.81±2.38	2.304E-02	3.25E+01
12.82±0.54	25.13±3.36	2.401E-02	3.19E+01
12.21±0.52	32.98±3.86	2.429E-02	3.15E+01
11.63±0.49	34.24±4.01	2.498E-02	3.07E+01
11.53±0.49	30.60±3.35	2.728E-02	2.97E+01
10.82±0.46	47.91±5.24	2.881E-02	2.87E+01
10.34±0.44	53.21±5.68	3.173E-02	2.76E+01
10.08±0.43	61.94±6.61	3.654E-02	2.63E+01
9.29±0.39	80.94±9.00	4.078E-02	2.48E+01
8.92±0.38	84.75±9.42	4.824E-02	2.29E+01
8.45±0.36	81.60±9.07	5.602E-02	2.06E+01
7.55±0.32	47.82±5.32	5.831E-02	1.78E+01

5-4 Differential, and Integral Yields for Mo, and Gd Targets

From the excitation function curves given in Figures from (5-1) to (5-14), the differential thin target yields for the production of ^{92m}Nb, ^{94}Tc, ^{95m}Tc, ^{95g}Tc, ^{96mg}Tc, and ^{99m}Tc via the ^{nat}Mo(p,x) reactions, and ^{152}Tb, ^{154m}Tb, ^{154g}Tb, ^{155}Tb, ^{156}Tb, and ^{160}Tb via the ^{nat}Gd(p,x) reactions, respectively have been calculated. The differential yields as shown in Figures (5-15), and (5-16) were calculated by assuming the beam current as 1 μA and the irradiation time as 1h. The calculated integral thick target yields as a function of target thickness (in MeV) are shown in Figures (5-17) to (5-28) for the production of ^{92m}Nb, ^{94g}Tc, ^{95g}Tc, ^{95m}Tc, ^{96mg}Tc, ^{99m}Tc, ^{152g}Tb, ^{154m}Tb, ^{154g}Tb, ^{155}Tb, ^{156}Tb, and ^{160}Tb, respectively.

The integral yields were calculated by assuming the incident beam current is collecting a charge of value one Coulomb. Whenever, both activities per microampere hour or per collected charge of one coulomb are interrelated. This is a rational quantity, where one can use it to produce a considerable amount of activity. Thin target yields deduced from the excitation fonction for ^{nat}Mo(p,x)^{92m}Nb, ^{94g}Tc, ^{95g}Tc, ^{95m}Tc, $^{96m+g}$Tc, and ^{99m}Tc processes. The yield of ^{92m}Nb, ^{95g}Tc, ^{95m}Tc, and $^{96m+g}$Tc is multiplied by factors as shown in Figure (5-15).

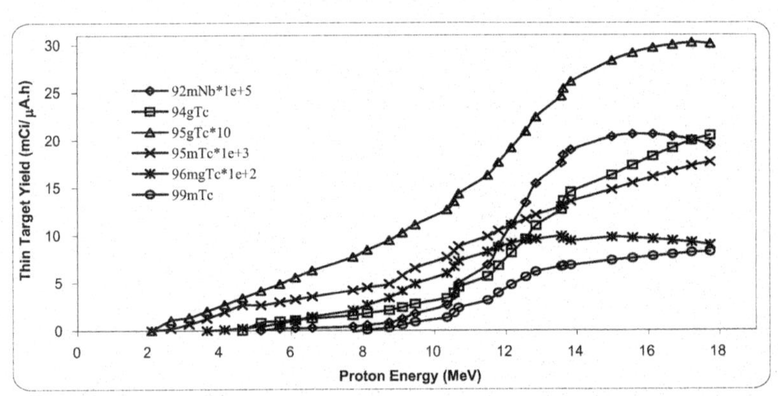

Figure (5-15): The Differential Thin Targets Yields for the Production of ^{92m}Nb, ^{94}Tc, ^{95m}Tc, ^{95g}Tc, ^{96mg}Tc, and ^{99m}Tc via the ^{nat}Mo(p,x) Reactions.

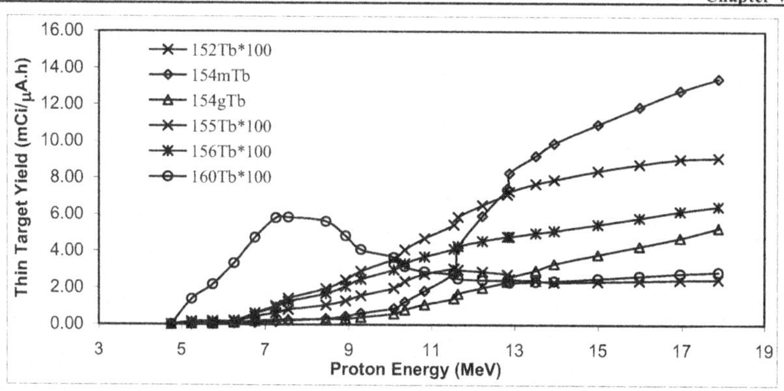

Figure (5-16): The Differential Thin Targets Yields for the Production of ^{152g}Tb, ^{154m}Tb, ^{154g}Tb, ^{155}Tb, ^{156}Tb, and ^{160}Tb via the ^{nat}Gd(p,x) Reactions.

Thin target yields deduced from the excitation fonction for ^{nat}Gd(p,x)^{152}Tb, ^{154m}Tb, ^{154g}Tb, ^{155}Tb, ^{156}Tb, ^{160}Tb processes. The yields of ^{152}Tb, ^{155}Tb, ^{156}Tb and ^{160}Tb are multiplied by 100. Thick target yield Y(E,ΔE) is defined as a two parameters function of incident energy E(MeV) on the target and energy loss in the target itself ΔE (MeV).the approximation of a monochromatic beam of energy E is hold, not affected by either intrinsic energy spread or straggling. In case of total particle energy absorption in the target (i.e. energy loss ΔE=E), the function Y(E,ΔE) reaches a value Y(E,E-E_{th}), for ΔE=E-E_{th}, that represent mathematically the envelope of the Y(E,ΔE) family of curves. This envelope is a monotonically increasing curve, never reaching either a maximum or a saturation value, even if its slope becomes negligible for high particle energies and energy losses. It is obviously that the production yield of a thick-target does not increase further, if the residual energy in the target is lower than the nuclear reaction energy threshold E_{th}. In practice, the use of a target thickness larger than the "effective" value, is unsuitable from technological point of view, due to the larger power density Pd(W/g) deposited by the beam in target material itself, instead of target cooling system.

The analytical expressions of thick-target yields y(E) were analytically and numerically integrated at 0.5 MeV intervals obtaining the families of curves reported in Figures (5-17), and (5-29). In the same pictures, the first and second calculated loci of the maxima of thick-target yield are represented (the second maximum is present only in some cases). These maxima correspond to couples of optimized values (E, ΔE), having different values for each different radionuclide.

This set of thick-target yields allows calculating the optimum irradiation conditions to have radiotracers with higher radionuclidic purity as possible. From the present cross section and yield data measurements we conclude that the optimum energy range for the production of selected isotope via the reactions induced by protons on natural ^{nat}Mo and ^{nat}Gd, the yield values for other ranges can be easily obtained from Figure (5-15), (5-16). Considering that some laboratories having a small-sized cyclotron with maximum proton energies of 16-18 MeV may decide to produce ^{99m}Tc locally via this route, It seems to be a feasible option, i.e. a medium energy cyclotron could be used.

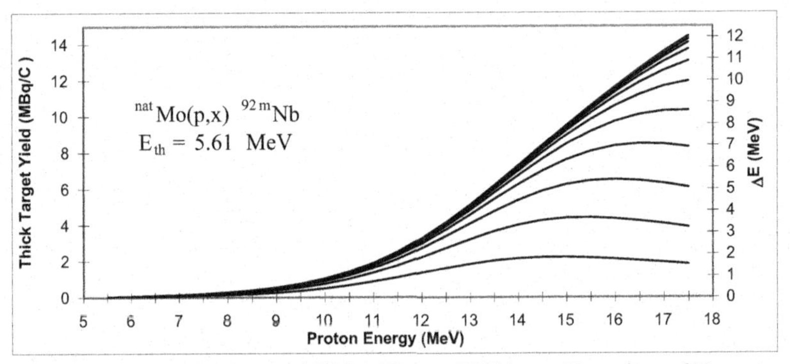

Figure (5-17): The Integral Thick Target Yield as a Function of Target Thickness (in MeV) for the Production of ^{92m}Nb.

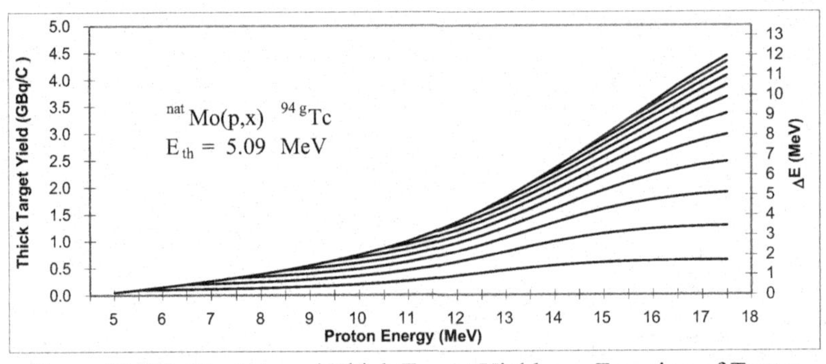

Figure (5-18): The Integral Thick Target Yield as a Function of Target Thickness (in MeV) for the Production of ^{94g}Tc.

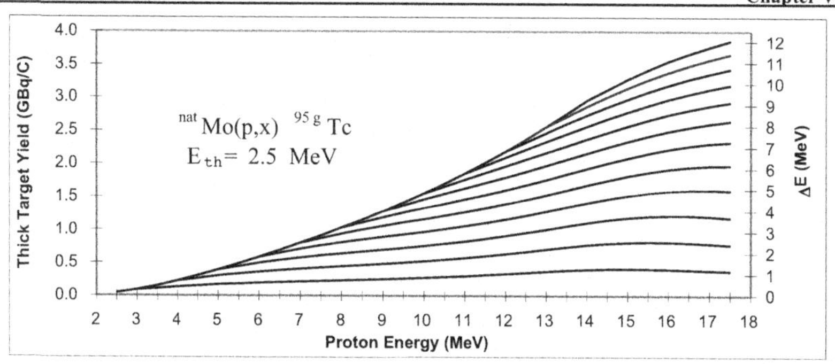

Figure (5-19): The Integral Thick Target Yield as a Function of Target Thickness (in MeV) for the Production of ^{95g}Tc.

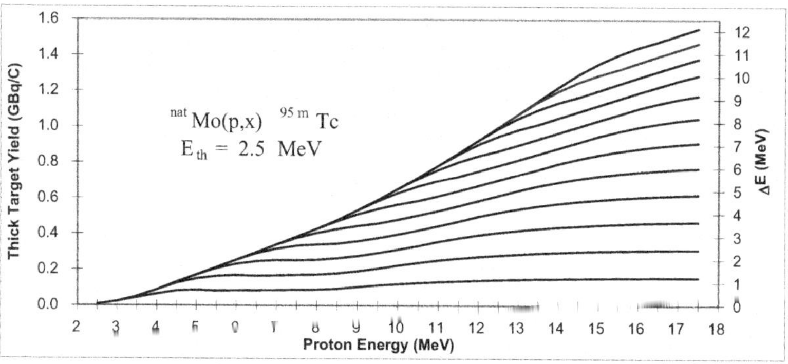

Figure (5-20): The Integral Thick Target Yield as a Function of Target Thickness (in MeV) for the Production of ^{95m}Tc.

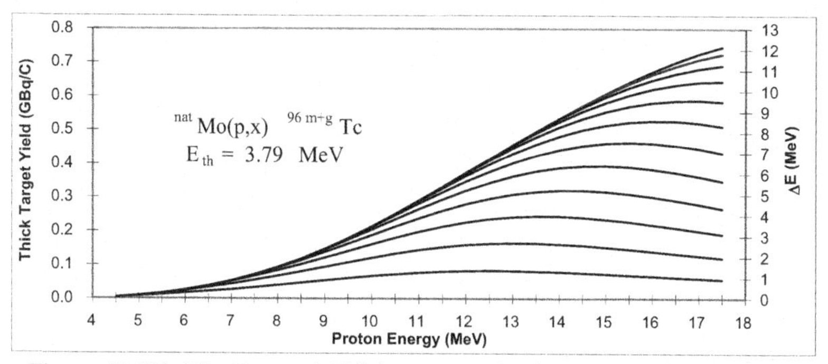

Figure (5-21): The Integral Thick Target Yield as a Function of Target Thickness (in MeV) for the Production of $^{96m+g}Tc$.

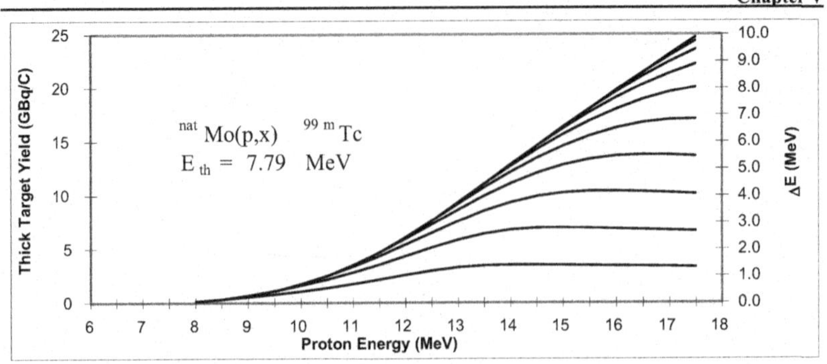

Figure (5-22): The Integral Thick Target Yield as a Function of Target Thickness (in MeV) for the Production of ^{99m}Tc.

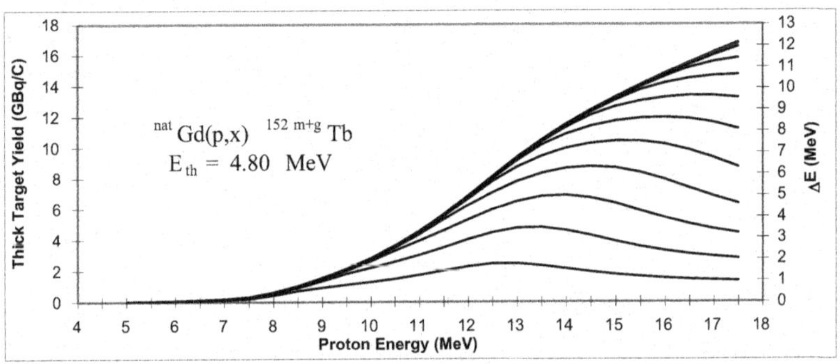

Figure (5-23): The Integral Thick Target Yield as a Function of Target Thickness (in MeV) for the Production of $^{152m+g}Tb$.

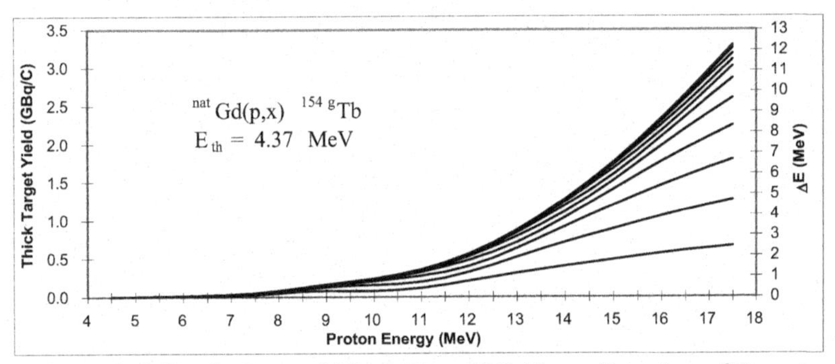

Figure (5-24): The Integral Thick Target Yield as a Function of Target Thickness (in MeV) for the Production of ^{154g}Tb.

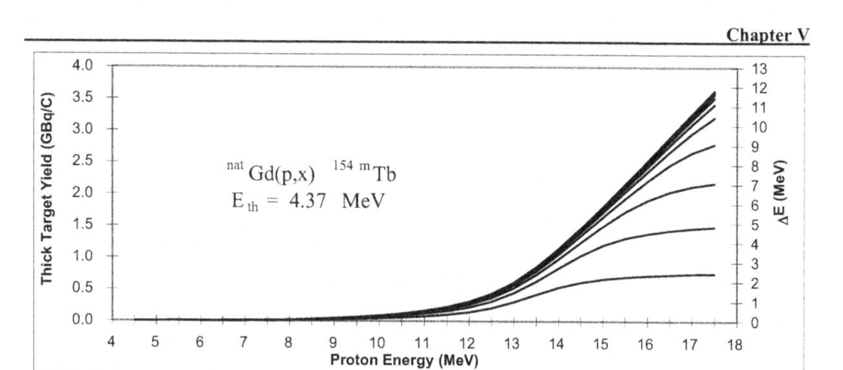

Figure (5-25): The Integral Thick Target Yield as a Function of Target Thickness (in MeV) for the Production of ^{154m}Tb.

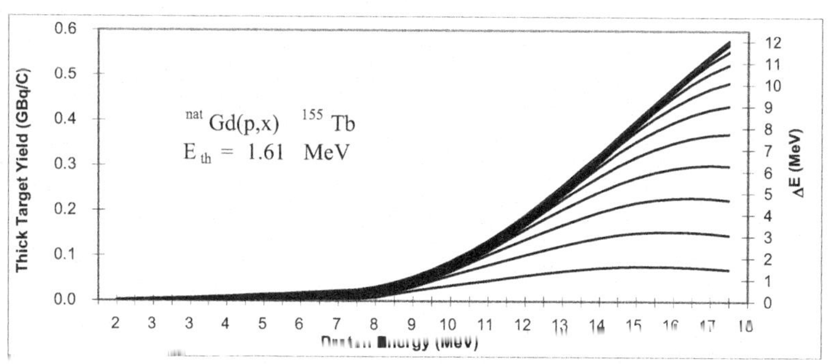

Figure (5-26): The Integral Thick Target Yield as a Function of Target Thickness (in MeV) for the Production of ^{155}Tb.

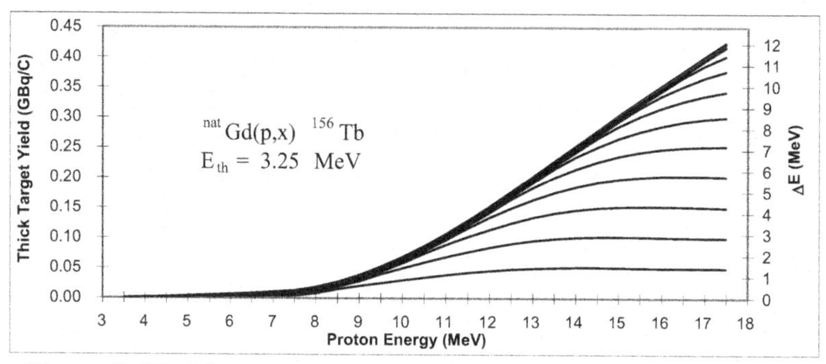

Figure (5-27): The Integral Thick Target Yield as a Function of Target Thickness (in MeV) for the Production of ^{156}Tb.

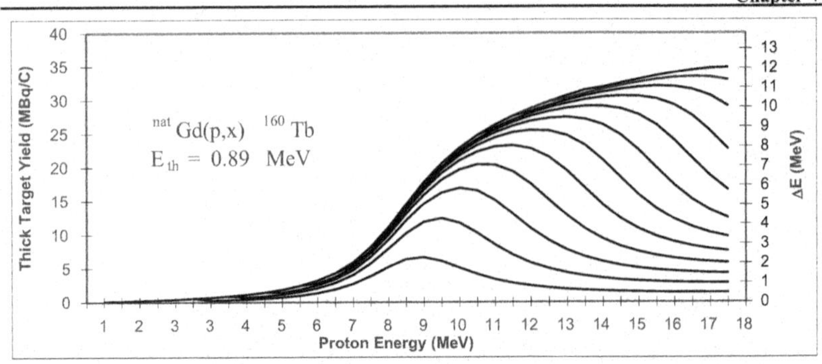

Figure (5-28): The Integral Thick Target Yield as a Function of Target Thickness (in MeV) for the Production of ^{160}Tb.

5-5 Calibration Curves for Thin Layer Activation Technique

For investigation of wear, corrosion or erosion the thin layer activation (TLA) technique is a very useful method. The technique is based, on the control of the removed surface layer through measurement of its radioactivity. The knowledge of the distribution of the radioactivity as a function of the depth (so-called, calibration curve) is essential. The radioactivity is generally produced directly in the investigated sample. Reliable cross-section data play important rule for determination of the calibration curves. The application of nuclear data accelerates significantly the determination of the calibration curves or to select the proper primary bombarding energy particle to assure constant, well known activity distribution in the investigated volume of the samples. Various applications require different activation profile, which can be fulfilled with different nuclear reactions, different bombarding beams and different, primary energies. In such a way "the induced reaction play important role in these applications. In most TLA studies on Cu the proton induced reactions play more important role, compared with other charged particle reactions. The availability of the proton is broader, and the yield of the ^{nat}Cu(p,x)^{65}Zn reaction is high. TLA is frequently used for copper, which is an important basic material of many tools and basic component of different alloys. The half-life of ^{65}Zn is usually used, the production yield of ^{65}Zn is in considerable amount, suitable for TLA studies. In Figure (5-29), we present the integral thick target yield for ^{nat}Mo(p,x)^{96}Tc reaction based on the present measurement in comparison with the other literature data reported by M. Bonardi et al., (2002); S. Takacs et al., (2002) [cf. **3**,**86**].

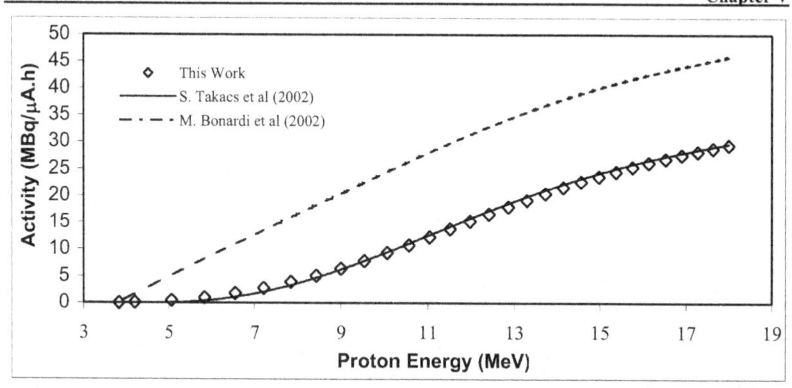

Figure (5-29): Presents the Integral Thick Target Yield for $^{nat}Mo(p,x)^{96}Tc$ Reaction in Comparison with the other Literature Data.

Comparing the three curves one can conclude that the yields calculated here are lower than the previously cited ones. Due to the suitable production and decay parameters of ^{96}Tc using the TLA method can effectively use it as tracer isotope for wears measurement of construction elements of different structures with Mo contents. The activity distribution as function of the penetrating depth of the proton beam is shown in Figure (5-30). The calibration curve calculated from our cross-section data is shown in Figure (5-30) in comparison with the calibration curve taken from the database of the M. Bonardi et al., (2002), S. Takács et al., (2002) [cf. **3**,**86**]. As it is seen in Figure (5-30), the comparison was made at three different proton-bombarding energies (18, 14 and 10 MeV). The curves exhibit some disagreement with M. Bonardi et al., (2002) [cf. **3**]. At low energy the solid curve calculated from our data is lower in amplitude, while at higher energy the ratio is opposite, which is not surprising taking into account the above mentioned method, and original database.

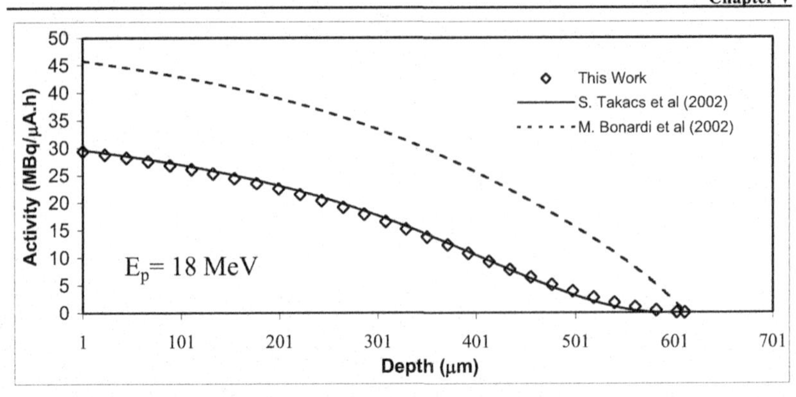

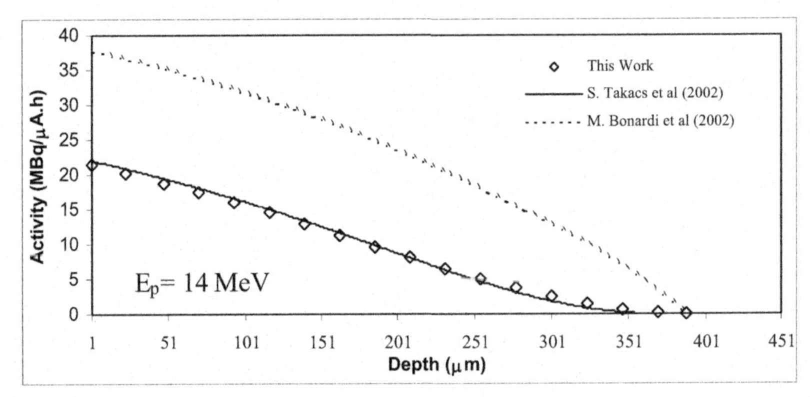

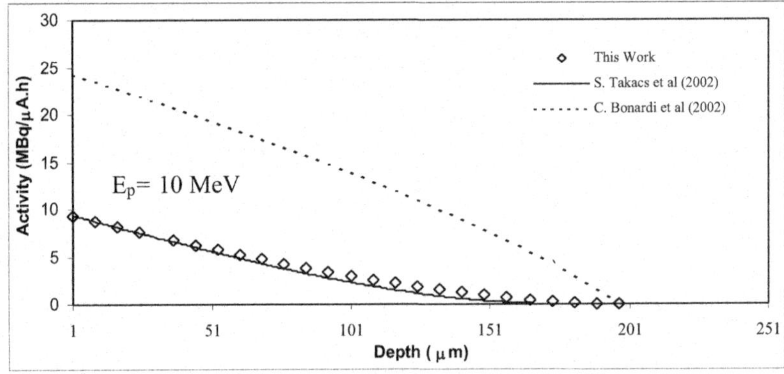

Figure (5-30): The Activity Distribution for $^{nat}Mo(p,x)^{96}Tc$ Reaction as a Function of the Penetrating Depth of the 18, 14, and 10 MeV Bombarding Proton Beam Calculated for Mo Using the Fitted Cross Sections of this Work in Comparison with other Literature Data.

5-6 Discussion

5-6.1 Influence of Secondary Particles

The study presented here includes measurement of most nuclear data induced by proton on natural Mo, Gd. All nuclear reactions considered are endoergic, with negative ***Q***-values, thus the calculated energy threshold for the different proton induced reactions Tables (2-3), (2-4) were also taken into account, imposing a yield value equal to zero.

While measuring the $^{nat}Mo(p,2n)^{99m}Te$ reaction cross section, a contribution via the $^{100}Mo(p,pn)^{99}Mo \rightarrow {}^{99m}Tc$ process to the count rate of ^{99m}Tc had to be subtracted. Two methods could possibly be used to apply this correction: (a) via a calculation using the fitted excitation curve of the $^{100}Mo(p,pn)^{99}Mo$ reaction. or (b) individual correction for each γ-ray measurement of ^{99m}Tc via the independently determined ^{100}Mo-activity (using the two above-mentioned γ-lines in the same sample). We decided to correct the count rates of ^{99m}Tc via the second method. This was more tedious but reduced the individual error caused by the uncertainties in the area weight of the sample and the detector efficiency.

While studying the $^{98}Mo(p,\gamma)^{99m}Tc$ reaction, account had to be taken of the two strong reactions on ^{100}Mo leading to the formation of ^{99m}Tc (i.e. $^{100}Mo(p,2n)^{99m}Tc$ and $^{100}Mo(p,pn)^{99}Mo \rightarrow {}^{99m}Tc$, one reaction leads to ^{90}Nb through $^{nat}Mo(p,x)$-process. However, a correction was also needed for the first count due to the contribution of the 141.2 keV (66.7%) photons from the decay of ^{90}Nb (14.6 h). After an elapsed time >170 h to allow for the decay of the directly produced ^{99m}Tc (6.02 h), and of the ^{90}Nb (14.6 h) radioactivities. A second radioassay was conducted to measure the direct production of the longer-lived 66 h ^{99}Mo parent. A correction in the count rate was therefore applied for the low $^{98}Mo(p,\gamma)^{99m}Tc$ content due to its very low cross section, using the measured cross sections of the two contributing processes, since the irradiation of natural Mo. The counts rate corrections had large errors since large numbers had to be subtracted from each other, it should be mentioned that the $^{98}Mo(p,\gamma)^{99m}Tc$ reaction, the observed count rates of ^{99m}Tc were anyway rather low. So we used the revealed data reported by Scholten et al., (1999) [cf. **44**] to correct its contribution.

It has to be considered, that in the ^{98}Mo sample secondary neutrons can also produce ^{99}Mo via the $^{100}Mo(n,2n)^{99}Mo$, $^{98}Mo(n,\gamma)^{99}Mo \rightarrow {}^{99m}Tc$, and $^{98}Mo(p,\gamma)^{99m}Tc$ reactions. However, no corrections were applied for some of these possible contributions due to the following three reasons:

1. The Q-value of the $^{100}Mo(n,2n)^{99}Mo$ reaction is highly negative (–8.0 MeV). The secondary neutron spectrum produced in the interaction of protons with the four elements encountered in each stack Ti, Al, Cu and Mo would have the maximum intensity at 2-4 MeV and a tail extending almost up to the maximum energy of the incident protons. However, since low proton beam currents were used and the stacks were rather thin, the intensity of the high energy neutrons would be very low and thus the contribution of the $^{100}Mo(n,2n)^{99}Mo$ process to the total formation of ^{99}Mo infinitesimally is small.
2. The cross section of the $^{98}Mo(p,\gamma)^{99m}Tc$ reaction in the MeV region is only a few mb as mentioned by McLane et al., (1988) [cf. **167**].
3. In highly enriched ^{98}Mo irradiated with protons carried out by Scholten et al., (1999) [cf. **44**], and the radioisotope ^{99}Mo was definitely observed. However, the calculated contribution from the $^{100}Mo(p,pn)^{99}Mo$ reaction on the small amount of ^{100}Mo present in the ^{98}Mo sample was very nearly equal to the measured ^{99}Mo activity as reported by Scholten et al., (1999) [cf. **44**].
4. The contribution arising from the reactions induced by secondary neutrons was thus, possibly originate from the difference in the case between the use of isotopically enriched or natural targets, where the production of neutron secondary particles, its built up is appreciable may induce another indirect reactions such as $^{98}Tc(n,\gamma)^{99m}Tc$. So we could not push it within the limits of experimental errors.

5-6.2 Trends in Integral Cross Sections

By simultaneous measurement of the excitation curves using multi monitor reactions giving a fair reproduction of the recommended cross section values in the whole energy range covered, in spite of it is true for some cases. We can not look at it as a global view. This implies many of deviations in the other cases, originated from many causes. The use of one or more monitor attached to sample let us must make a comprehensive understand for the relation could be generate between both, in their response to the incident beam, profile of excitation functions belong to each one for monitor (known) for sample (unknown). The change in thresholds between both in the interest energy range. For approximation, looking at monitor as an infinitesimal laminar could not affect any interactions carried around originate from itself, it represent only a detector. But, we can not extreme the effect of the surroundings on it that could be change its detect value.

In some cases the sample is sensitive to the primary beam as well as ejected secondary beam. As we recognized in molybdenum, we strongly believe that our cross-section data measured on Mo have the same reliability as that measured on enriched samples by taking into account the effect of the secondary beam that elevate the output results. Therefore, results were checked for the multi-monitor reactions we used in this work and the data calculated with inter-consistent flux were found to be higher than the recommended values by 12-26% (in average 20%) around the maximum of the excitation function. Beside of using multi monitors to detect the primary incident beam we generate the effective incident beam value through correction operator for flux, considering the neutron interaction properties of the sample introduced using experimental results or using theoretical model codes. That giving a glance for the neutron production, transmission and the value added to the currently incident proton beam.

The short half-life of the ^{63}Zn isotope could lead to high sample activity, overloading the detector system or improper dead time correction and giving too low activity values. Also a slight energy shift can be observed for the three-excitation functions toward lower energies. Accepting the average with a good agreement between the published data set and our results. Detailed analyses were made on the most frequently used monitor reactions for proton beams, among them are the ^{nat}Ti(p,x)^{48}V, ^{nat}Cu(p,x)^{63}Zn, and ^{nat}Cu(p,x)^{65}Zn processes.

The cross-section of the ^{96}Nb producing process is however small. Improper waiting time or an incomplete separation of its contribution in calculating ^{96mg}Tc could cause only minor error in the resulting cross-section of the ^{96mg}Tc process.

We used ALICE-91, and EMPIRE-II (statistical and pre-equilibrium model) to predict the production of ^{99}Mo by proton induce reaction from ^{100}Mo. The codes were used to calculate the excitation functions for proton induced reactions on ^{100}Mo that led to the production of Nb, Mo and Tc isotopes. The codes predicts a maximum cross section of the order of 100 mb for the ^{100}Mo(p,pn)^{99}Mo nuclear reaction. The ALICE-91 code cannot be used to predict the cross sections for the isomeric states for the ^{100}Mo(p, 2p)$^{99m+g}$Nb$\rightarrow^{99}$Mo reactions. However, the predicted cross section for protons on ^{100}Mo to yield ^{99}Nb suggests that the (p,2p) reaction channel does not contribute significantly to the thick target yield of ^{99}Mo.

5-6.3 Prospective for Fluence Response Modeling

The treatment advanced in this work is built on a number of assumptions. For the sake of providing a focus the main assumptions are summarized below.

1. Proton interaction is predominantly Coulomb scattering.

2. The recoil neutrons for a given proton energy, E_p, are mono-energetic. (The value of the neutron energy, E, is taken to be equal to the maximum-kinetic energy the recoil neutron can possess). Will suppose also the generated neutron will have nearly the same energy and vector motion as the primitive beam generate it. So will consider only for succeeding processes after generation.

3. The cross section for the production of neutrons used in these calculations is a certain energy-weighted component of the production cross section. It also enables us to use a pseudo-recoil neutron beam parallel to the incident proton beam, which simplifies the treatment.

4. The integration or takings into account the sum of two contributing formulae for discrete energetic protons and (wide) range neutrons, also for single target foil and subsequent successive foils constitute the stacked foils.

5. We faced with an angular deviation due to a specific account is suffered by a neutron due to multiple and plural scattering through the target material. Our deal now arranged to determine the effective flux through probability response function. The deviation from the straight-ahead direction due to the scattering of recoil neutrons from the sample atoms is assumed to takes place along one direction, namely that corresponding to the mean angle of scattering in the limit of large numbers of collisions.

6. The assumption that the approximately linear relations between incident projectile energy using certain target foil, the (p,n) cross section results remaining essentially constant with energy is invalid. The energy dependence of this cross section over the energy range of interest is essential. So, integration over the energy range involved in calculation over all energy range. The net response in turn will be dependent on the energy.

7. The energy spread in the incident beam will be neglected. All relevant quantities to head-on projections for incident proton, which we shall call the axial approximation. The effective region of the sample for a given E_p is bounded by lines originating from points on the sample axis. The mean scattering angle and the range in sample give the direction and length of the line. Edge effects are neglected in that the volume of the effective region is taken to be independent of the location at which the recoil neutron is produced.

8. We gave a systematic development of the probability function relating to the production process, transmission of targeted neutrons.

We can conclude from these assumptions that, we bind some of correcting errors to be originated from all counting techniques. Those retrieve the correction factor as a shift in the overall efficiencies of the detecting system. To attain what so called, effective efficiency, now by introducing the theory of fluence response model as a convenient tool to describe the flux through the stacked foils. It seem to be representing an efficient method, we can condsider it as a proportional constant that connect the measured activity to measured flux. This constant represent the probability response function for all processes contribute in it. So the over all correction seem to be as a multiplying of all the probabilities, the product is the proportional constant of the detection efficiency.

The extent and validity of these assumptions have been discussed in the appropriate previous sections. The refinement or rather the elimination of most of these assumptions must form the next stage of the fluence response function calculations. Such a modification is expected to increase the absolute flux accuracy, finally the measured cross sections.

5-7 Conclusion

1. Up to our best knowledge the obtained data for Gd are the first. All nuclear reactions considered are endoergic, with negative *Q*-values.

2. The contributions in the measuring reaction cross section due to secondary particle processes to the count rate have to be considered.

3. Two strong reactions on ^{100}Mo leading to the formation of ^{99m}Tc (i.e. $^{100}Mo(p,2n)^{99m}Tc$, $^{100}Mo(p,pn)^{99}Mo \rightarrow {}^{99m}Tc$), and one reaction leads to ^{90}Nb through $^{nat}Mo(p,x)$-process which gives the same γ-lines.

4. A correction was also needed for the first count due to the contribution of the 141.2 keV (66.7%) photons from the decay of ^{90}Nb (14.6 h). After an elapsed time >170 h to allow for the decay of the directly produced ^{99m}Tc (6.02 h), and of the ^{90}Nb (14.6 h) radioactivities.

5. Secondary neutrons could produce ^{99m}Tc via the ^{100}Mo(n,2n)^{99}Mo$\rightarrow$^{99m}Tc, ^{98}Mo(p,γ)^{99m}Tc. However, no possible contribution due to the following reasons. The Q-value of the ^{100}Mo(n,2n)^{99}Mo reaction is highly negative (–8.0 MeV). The cross section of the ^{98}Mo(p,γ)^{99m}Tc reaction in the MeV region is only a few mb as mentioned by McLane et al., (1988).

6. Multi-monitor reactions giving a fair reproduction of the recommended values in the whole energy range covered.

7. ALICE-91, and EMPIRE-II were used to calculate the excitation functions for proton induced reactions on ^{nat}Mo, and ^{nat}Gd. The EMPIRE-II code could be used to predict the cross sections for the isomeric states. However, ALICE-91 not.

8. The response function probability for all processes contributing in radionuclide production is a convenient tool to account the elevation in the measured activity.

References:

1- International Atomic Energy Agency, Vienna, Austria. TECDOC-1211, (2001).

2- Pierre Marmier, Eric Sheldon, "Physics of Nuclei and Particles", Laboratory of Nuclear Physics, Vol. I, Federal Institute of Technology, Zurich, Switzerland, Publisher Academic Press, New York, (1969).

3- Mauro Bonardi, Claudio Birattari, Flavia Groppi, Enrico Sabbioni, "Thin-Target Excitation Functions, Cross-Sections and Optimized Thick-Target Yields for $^{nat}Mo(p,xn)^{94g,95m,95g,96(m+g)}Tc$ Nuclear Reactions Induced by Protons from Threshold up to 44 MeV. No Carrier Added Radiochemical Separation and Quality Control", Appl. Radiat., and Isot. 57, 617–635, (2002).

4- R.B. Firestone, C.M. Baglin, F.S.Y. Chu, and J. Zipkin, "Table of Isotopes", 8th Edition, Vols. I and II. Wiley, New York, (1996); R.B. Firestone, C.M. Baglin, and F.S.Y. Chu, "Table of Isotopes", 8th Edition. Update on CD-ROM. Wiley, New York, (1998).

5- S.M. Qaim, in the proceeding of the Second Conference and Workshop on Cyclotrons and Applications, Cairo, EGYPT, p.p. 202, March (1997)

6- S. M. Qaim, Institut für Chemie 1 (Nukiearchemie), Kernforschungsanlage Julich GmbH, D-5170 Julich, Federal Republic of Germany, "Nuclear Data Relevant to Cyclotron Produced Short-Lived Medical Radioisotopes", Radiochimica Acta 30, 147-162 (1982).

7- L.M. Freeman (ed.), Freeman and Johnson's "Clinical Radionuclide Imaging", 3rd edition., Grune & Stratton, USA, (1984).

8- H. Vera Ruiz, Report of an International Atomic Energy Agency's Consultants' Meeting on Fluorine 18: Reactor Production and Utilization, Appl. Radiat. Isot. 39, 31-39, (1988); H. Vera Ruiz (ed.), "Summary Report on the Research Co-ordination Meeting of the Co-ordinated Research Project on "Standardized High Current Solid Targets for Cyclotron Production of Diagnosis and Therapeutic Radionuclides", Brussels, Belgium, 27-30 Sep., (2000).

9- D.J. Silvester, "Radioisotope Production: An Historical Introduction, In Radionuclides Production", vol. I (Eds. F. Helus and L. G. Colombetti), CRC Press, Inc., Florida, (1983).

10- H.O. Anger, "Scintillation Camera", Rev. Sci. Instr., 29, 27-33, (1958).

11- R.E. Boyd, "Molybdenum-99/Technetium-99m Generator", Radiochim. Acta 30, 123-145, (1982).

12- R.E. Boyd, "The Special Position of ^{99m}Tc in Nuclear Medicine, In Radionuclides Production", vol. II (Eds. F. Helus and L. G. Colombetti), CRC Press, Inc., Florida (1983).

13- P.O. Denzler, P. Rosch, and S.M. Qaim, "Excitation Functions of α-Particle Induced Nuclear Reactions on Highly Enriched ^{92}Mo Comparative

Evaluation of Production Routes for 99mTc". Radiochimica Acta. 68: 13-20, (1995).

14- O. Solin, Ph.D. Thesis, "Interaction of High Energy Charged Particles with Gases", Acta Academia Aboensis, Ser. B, Vol., 48, No. 2, (1988).

15- S.M. Qaim, A. Hohn, Th. Bastian, K.M. El-Azoney, G. Blessing, S. Spellerberg, B. Scholten, H.H. Coenen, "Some Optimization Studies Relevant to the Production of High-Purity ^{124}I and ^{122g}I at a Small-Sized Cyclotron", Appl. Radiat. Isot. 58, 69-78, (2003).

16- J. Solin, A. Bergman, M. Haaparanta and A. Reissell, "Production of ^{18}F from Water Targets. Specific Radioactivity and Anionic Contaminants". Appl. Radiat. Isot. 39, 1065-1071, (1988).

17- S.M. Qaim, "Target Development for Medical Radioisotope Production at a Cyclotron", Nucl. Instr. and Meth. A 282, 289-295 (1989); S. Zeister, and F. Helus (Eds), Proceedings of the Seventh Workshop on Targetry and Target Chemistry (WTTC97), DKFZ Press Dept. Heidelberg, Germany, June 8-11, (1987).

18- S. Spellerberg, P. Reimer, G. Blessing, H.H. Coenen, S.M. Qaim, "Production of ^{55}Co and ^{57}Co via Proton Induced Reactions on Highly Enriched ^{58}Ni" Appl. Radiat. Isot. 49, 1519-1522, (1998).

19- K. Gul, A. Hermanne, M.G. Mustafa, F.M. Nortier, P. Oblozinsky, S.M.Qaim, B. Scholten, Yu. Shubin, S. Takacs, T.F. Tarkanyi, Y. Zhuang, "Charged-Particle Cross Section Database for Medical Radioisotope Production" IAEA-TECDOC-1211, IAEA, Vienna, Austria (2001).

20- Donald J. Hughes, "Neutron Cross Section", Brookhaven National Laboratory, and Published by Pergamon Press, London, (1957).

21- Moazzem Hossein Maih, Jochen Kuhnhenn, Ulrich Herpers, Rolf Michel, Peter Kubik, "Production of Residual Nuclides by Proton Induced Reactions on Target W at an Energy of 72 MeV", J. of Nucl. Sci. Tech., Supplement 2, p. 369-372, (2002).

22- R. Michel, R. Bodemann, H. Busemann, R. Daunke, M. Gloris, H. J. Lange, B. Klug, A. Krins, I. Leya, M. Lupke, S. Neumann, H. Reinhardt, M. Schhatz-Buttgen, U. Herpers, Th. Schiekel, F. Subdrock, B. Holmqvist, H. Conde, P. Malmborg, M. Suter, B. Dittrich-Hannen, P. W. Kubik, and H. A. Synal. Nucl. Instr. and Meth. B, 129, 153, (1997).

23- S.M. Qaim, "Activation Cross Sections, Isomeric Cross Sections Ratios and Systematics of (n,2n) Reactions at 14-15 MeV". Nucl. Phys. A185:614-624, (1972)..

24- Y. Nagame, S. Baba, and T. Saito, "Isomeric Yield Ratios for the ^{95}Mo(p,n)^{95m,g}Tc Reaction", Appl. Radiat. Isot., Vol. 45, No. 3, pp. 281-285, (1994).

25- P.G. Young, and E.D. Arthur, "GNASH: Pre-Equilibrium Statistical Nuclear Model Code for Calculations of Cross Sections and Emission Spectra". Report No. LA-6947. Los Alamos Scientific Laboratory, (1977).

26- M. Bonardi, "The contribution to nuclear data for biomedical radioisotope production from the Milan Cyclotron Laboratory", Data Requirements for Medical Radioisotope Production, Proc. Consultants Mtg Tokyo, Japan, 1987 (K., Okamoto ed.), Rep. INDC(NDS)-193, IAEA, Vienna (1988).

27- J.F. Ziegler, "SRIM 2003 code, SRIM.com, 1201 Dixona Dr., Edgewater, 21037, USA, available from www.SRIM.org, (2003).

28- P.P. Dmitriev, "Radionuclide Yield in Reactions with Protons, Deuterons, Alpha Particles and Helium-3" (Handbook), Moscow, Energoatomizdat, 1986; Rep. INDC(CCP)-263, IAEA, Vienna (1986).

29- M. Blann, NEA Data bank, France, Report PSR-146, (1991); M. Blann, "Calculation of Excitation Functions with Code ALICE". In K. Okamoto, (Ed.), INDC(NDS)-195/GZ. International Atomic Energy Agency, Vienna, (1988).

30- M. Herman, "EMPIRE-II Statistical Model Code for Nuclear Reaction Calculations", Version 2.19 (Londi), IAEA, Vienna, Austria, March, (2005). Available from: http://www-nds.iaea.org/empire/, (2005).

31- D. Tucker, M.W. Greene, A.J. Weiss, A. Murrenhoff, "Methods of Preparation of some Carrier-Free Radioisotopes Involving Sorption on Alumina". Report BNL-3746, BNL, Upton, NY, (1958).

32- R. Powell, J. Steigman, Technetium. In: G. Subramanian, B.A. Rhodes, J.F. Cooper, V.J. Sodd, (Eds.), "Radiopharmaceuticals". The Society of Nuclear Medicine Inc., USA, pp. 21-70, (1975).

33- R.E. Boyd, "The Special Position of ^{99g}Tc in Nuclear Medicine". In: F. Helus, (Ed.), Radionuclides Production, Vol., II, CRC Press, Boca Raton, FL, pp. 125-152, (1983).

34- R.J. Nickles, B.T. Christian, C.C. Martin, A.D. Nunn, C.K. Stone, "Tc-94m Radionuclidic Purity Requirements for Pharmacokinetic Studies with PET". J. Nucl. Med. 32, 850, (1991).

35- R.J. Nickles, A.D. Nunn, C.K. Stone, S.B. Perlman, R.L. Levine, "Tc-94m Flow Agents: Bridging PET and SPECT". J. Nucl. Med. 32, 925, (1991).

36- R.J. Nickles, B.T. Christman, A.D. Nunn, C.K. Stone, "Cyclotron Production of High-Purity Tc-94m by in situ Sublimation". Proceedings of the 9th ISRC, Paris, pp. 447-448, (1992).

37- M. Sajjad, and R.M. Lambrecht, "Cyclotron Production of Medical Radionuclides". Nucl. Instr. Meth. B 79, 911-915, (1993).

38- M. Faβbender, A.F. Novgorodov, F. Roesch, and S.M. Qaim, "Excitation Functions of $^{93}Nb(^{3}He,xn)^{93m,g,94m,g,95m,g}Tc$-Processes from Threshold up to 35 MeV: Possibility of Production of ^{94m}Tc in High Radiochemical Purity Using a Thermo-Chromatographic Separation Technique. Radiochim. Acta 65, 215-221, (1994).

39- F. Roesch, and S.M. Qaim, "Nuclear Data Relevant to the Production of the Positron Emitting Technetium Isotope ^{94m}Tc via the ^{94}Mo(p,n)-Reaction". Radiochim. Acta 62, 115-121, (1993).
40- F. Roesch, A.F. Novgorodov, and S.M. Qaim. "Thermo-Chromatographic separation of ^{94m}Tc from Enriched Molybdenum Targets and its Large Scale Production for Nuclear Medical Application". Radiochim. Acta 64, 113-120, (1994).
41- F.O. Denzler, F. Roesch, and S.M. Qaim, "Excitation Functions of α-Particle Induced Nuclear Reactions on Highly Enriched ^{92}Mo: Comparative Evaluation of Production Routes for ^{94m}Tc". Radiochim. Acta 68, 13-20, (1995).
42- J. Rojas-Burke, "The Future Supply of Molybdenum-99". Newsline J. Nucl. Med. 36, 15N, (1995).
43- ALASBIN. XI Congress of the Latinamerican Association of Biology and Nuclear Medicine (ALASBIN). Santiago, Chile October 8-11, (1989).
44- Bernhard Scholten, Richard M. Lambrecht, Michel Cogneau, Hernan Vera Ruiz, and Syed M. Qaim, "Excitation Functions for the Cyclotron Production of ^{99m}Tc and ^{99}Mo" Appl. Radiat. Isot., 51, 69-80, (1999).
45- "Reaction Q-values, and Thresholds", Los Alamos National Laboratory, T-2 Nuclear Information Service. Available from <http://t2.lanl.gov/data/qtool.html>, (2005).
46- J.E. Beaver, H.B. Hupf, "Production of ^{99m}Tc on a Medical Cyclotron: a Feasibility Study". J. Nucl. Med. 12, 739, (1971).
47- G.L. Almeida, F. Helus, "On the Production of ^{99}Mo and ^{99m}Tc by Cyclotron". Radiochem. Radioanal. Lett. 28, 205, (1977).
48- R.M. Lambrecht, S.L. Waters, H. Lu, S.M. Qaim, H. Umezawa, G.J. Beyer, H. Heinzl, M. Bonardi, T. Nozaki, A. Hashizume, M.C. Lagunas-Solar, K. Kitao, and D. Berenyi, "Summary of Conclusions and Recommendations of Working Group I: Experimental Data. In: Okamoto, K. (Ed.), Consultants' Meeting on Data Requirements for Medical Radioisotope Production, Tokyo, Japan, (1987). Report INDC (NDS)-195/GZ. IAEA, Vienna, pp. 13-16, (1988).
49- M.C. Lagunas-Solar, P.M. Kiefer, O.F. Carvacho, C.A. Lagunas, Ya Po Cha, "Cyclotron Production of NCA ^{99m}Tc and ^{99}Mo. An Alternative Non-Reactor Supply Source of Instant ^{99m}Tc, and ^{99}Mo→^{99m}Tc Generators". Appl. Radiat. Isot. 42, 643, (1991).
50- M.C. Lagunas-Solar, "Production of ^{99m}Tc and ^{99}Mo for Nuclear Medicine Applications via Accelerators as an Option to Reactor Methods". Presented at the 18th Annual Conference of the Australian Radiation Protection Society. University of Sydney, Sydney, NSW, Australia, Oct. 6-8, (1993).

51- N. Levkowski, "Middle Mass Nuclides (A=40-100) Activation Cross Sections by Medium Energy (E=10-50 MeV) Protons and α-Particles (Experiment and Systematics)'. Inter-Vesti, Moscow, pp. 215, (1991).

52- H. Schopper, (Ed.), In: O. Madelung, (Ed.), Landolt Bornstein, Numerical Data and Functional Relationships in Science and Technology, Group I: Nuclear and Particle Physics, vol. 13: "Production of Radionuclides at Intermediate Energies, Sub-vol. (b): "Interactions of Protons with Targets from Kr to Te". Springer Verlag, Berlin, pp. 286, (1993).

53- H.P. Graf, and H. Muenzel, "Excitation Functions for α- Particle Reactions with Molybdenum Isotopes". J. Inorg. Nucl. Chem. 36, 3647-3657, (1974).

54- Z. Randa, and K. Svoboda, "Excitation Functions and Yields of (d,n) and (d,2n) Reactions on Natural Molybdenum". J. Inorg. Nucl. Chem. 38, 2289-2295, (1976).

55- I.N. Vishnevskil, V.A. Zheltonozhskil, and T.N. Lashko, "Measurement of the Isomeric Ratios for the Isotopes ^{93,94}Tc in Reactions Induced by Protons and α-Particles". Sov. J. Phys. 41, 910-912 (English translation from Yad. Fiz. 41, 1435-1439, (1985).

56- R.M. Lambrecht, and S.M. Montner, "Production and Radiochemical Separation of ^{92}Tc and ^{93}Tc for PET". J. Labeled Compd. Radiopharm. 19, 1434-1435, (1982).

57- R. Finn, T. Boothe, J. Sinnreich, E. Tavano, A. Gilson, and A.P. Wolf "Ancillary Cyclotron Production of Technetium-93m for Clinical and Chemical Research". Radiopharmaceuticals and Labeled Compounds 1984, (1985); IAEA-CN-45/22, IAEA, Vienna, Austria, pp. 47-54, (1985).

58- M.C. Lagunas-Solar, R.P. Haff, "Theoretical and Experimental Excitation Functions for Proton Induced Nuclear Reactions on Z=10 to Z=82 Target Nuclides". Radiochim. Acta 60, 57-67, (1993).

59- J.J. Hogan, "^{96}Mo(p,xn) Reaction from 10 to 80 MeV". Phys. Rev. C 6, 810-816, (1972).

60- J.J. Hogan, "Isomer Ratios of Tc Isotopes Produced in 10-65 MeV Bombardments of ^{96}Mo". J. Inorg. Nucl. Chem. 35, 705-712, (1973).

61- J.J. Hogan, "Comparison of the ^{94}Mo(p,n)^{94m,g}Tc and ^{96}Mo(p,n)^{96m,g}Tc Reactions". J. Inorg. Nucl. Chem. 35, 2123-2125, (1973).

62- E.A. Shakun, V.S. Batii, Yu.N. Rakivnenko, and O.A. Rastrepin, "Excitation Functions and Isomeric Ratios for the Interaction of Protons of Less than 9 MeV with Zr and, Mo Isotopes". Sov. J. Nucl. Phys. 46, 17-24 (English translation from Yad. Fiz. 46, 28-39, (1987).

63- M. Izumo, H. Matsuoka, T. Sorita, Y. Nagame, T. Sekine, K. Hata, and S. Baba, "Production of ^{95m}Tc with Proton Bombardment of ^{95}Mo". Appl. Radiat. Isot. 42, 297-301, (1991).

64- L. Lakosi, J. Safar, A. Vertes, T. Sekine, and K. Yoshihara, "Photonuclear Reactions of ^{99}Tc, Isomer Excitation Functions and Deexcitation, Implications in Nucleosynthesis". Radiochim. Acta 63, 23-28, (1993).
65- T. Sekine, K. Yoshihara, J. Safar, L. Lakosi, A. Vertes. "^{95}Tc, and ^{96}Tc Production by (γ,xn) Reactions". J. Radioanal Nucl. Chem. 186, 165-174, (1994).
66- N.G. Zaitseva, E. Rurarz, M. Vobecky, Hwan Kim Hyn, K. Novak, T. Tethal, V.A. Khalkin, and L.M. Popinenkova, "Excitation Function, and Yield for ^{97}Ru Production in ^{99}Tc(p,3n)^{97}Ru Reaction in 20-100 MeV Proton Energy Range". Radiochim. Acta 56, 59-68, (1992).
67- J.M.L. Ouellet, K. Oxorn, L.A. Hamel, L. Lessard, and C. Matte, "Cross Sections and Activation Profiles for Wear Monitoring". Nucl. Instr. Meth. B 79, 579-581, (1993).
68- R. B. Lauffer, "Paramagnetic Metal Complexes as Water Proton Relaxation Agents for MRI Imaging: Theory and Design". Chem. Rev. 87, 901, (1987).
69- S. C. Atlas, and R. C. Brasch, "Gadolinium Contrast Agents in Neuro-MRI". J. Comput. Assist. Tomogr. 17, Suppl. 1, (1993).
70- M. P. Lowe, "MRI Contrast Agents: The Next Generation", Aust. J. Chem., 55, 551, (2002).
71- D. Messeri, M. P. Lowe, D. Parker, and M. Botta, "A Stable, High Relaxivity, Diaqua Gadolinium Complex that Suppresses Anion and Protein Binding", Chem. Commun., 2742, (2001).
72- M. P. Lowe, D. Parker, O. Reany, S. Aime, M. Botta, G. Castellano, E. Gianolio, and R. Pagliarin, "pH Dependent Modulation of Relaxivity and Luminescence in Macrocyclic Gadolinium and Europium Complexes Based on Reversible Intramolecular Sulfonamide Ligation", J. Am. Chem. Soc., 123, 7601, (2001).
73- S. Blair, M. P. Lowe, C. E. Mathieu, D. Parker, P. K. Senanayake, and R. Kataky, "Optical pH Sensors Based on Luminescent Europium and Terbium Complexes Immobilized in a Soil Gel Glass", Inorg. Chem., 40, 5860, (2001).
74- J. I. Bruce, M. P. Lowe, and D. Parker, "Chapter 11 - Luminescence of Lanthanide (III) Chelates" in "The Chemistry of Contrast Agents in MRI", ed. A. E. Merbach, and E. Tóth, John Wiley & Sons, Ltd., Chichester, (2001).
75- J. I. Bruce, R. S. Dickins, L. Govenlock, T. Gunnlaugsson, S. Lopinski, M. P. Lowe, D. Parker, R. D. Peacock, J. J. P. Perry, S. Aime, and M. Botta. "The Selectivity of Reversible Oxy-Anion Binding in Aqueous Solution at a Chiral Europium, and Terbium Center.: Signalling of Hydrogencarbonate Chelation by Changes in the Form and Polarization of

Luminescence Emission Following Excitation at 365 nm", J. Am. Chem. Soc., 122, 9674, (2000).

76- S. Aime, E. Gianolio, E. Terreno, G. B. Giovenzana, R. Pagliarin, M. Sisti, G. Palmisano, M. Botta, M. P. Lowe, D. P. Parker, "Ternary Gd(III)LHSA Adducts: Evidence for the Replacement of Inner Sphere Water Molecules by Coordinating Groups of the Protein. Implications for the Design of Contrast Agents for MRI", J. Biol. Inorg. Chem., 5, 488, (2000).

77- M. P. Lowe, and D. P. Parker, "Controllable pH Modulation of Lanthanide Luminescence by Intramolecular Switching of Hydration State", Chem. Commun., 707, (2000).

78- C. Birattari, E. Gadioli, E. Gadioli Erba, A.M. Grassi Strini, G. Strini, G. Tadiaferri, "Pre-Equilibrium Processes in (p,n) Reactions", J., Nucl. Phys. A, 201, 579, (1973).

79- H. Piel, S.M. Qaim, and G. Stőcklin, "Excitation Functions of (p,xn)-Reactions on ^{nat}Ni and Highly Enriched ^{62}Ni: Possibility of Production of Medically Important Radioisotope ^{62}Cu at a Small Cyclotron". Radiochim. Acta 57, 1, (1992).

80- B. Scholten, Z. Kovacs, F. Tarkanyi, and S.M. Qaim,. "Excitation Functions of $^{124}Te(p,xn)^{124,123}I$ Reactions from 6 to 31 MeV with Special Reference to the Production of ^{124}I at a Small Cyclotron". Appl. Radiat. Isot. 46, 255, (1995).

81- E.A. Skakun, V.S. Batij, Y.U.N. Rakivnenko, O.A. Rasrtrepin, Isot, Sov. J. Nucl. Phys. 46, 17, (1987).

82- V.N. Levkowski, "Cross Section of Medium Mass Nuclide Activation (A=40-100) by Medium Energy Protons and Alpha Particles", Inter-Vesi, Moscow, USSR, (1991).

83- Yu. Zuravlev, P.P. Zarubin, A.A. Kolozvari, Izv. Acad. Nauk. Ser. Phys. 58, 106, (1994).

84- M.C. Lagunas-Solar, IAEA-TECDOC-1065, IAEA, Vienna, Austria, p.87, (1999).

85- Z. Wenrong, Yu. Weixiang, H. Xiaogang, L.U. Hanlin, "Excitation Functions of Reactions from d+Ti, d+Mo, p+Ti, and p+Mo". INDC(CRP)-044, p. 17, (1998).

86- S. Takacs, F. Tarkanyi, M. Sonck, and A. Hermanne, "Investigation of the $^{nat}Mo(p,x)^{96mg}Tc$ Nuclear Reaction to Monitor Proton Beams: New Measurements and Consequences on the Earlier Reported Data", Nucl. Instr. and Meth., 198, p.p. 183-196, (2002).

87- Directory of Cyclotrons, IAEA-DCRP/CD, (2005).

88- M.N.H. Comsan, "Status Report of Inshas Ion Accelerators", in the Proceeding of the Fifth Conference and Workshop on Cyclotrons and Applications, Cairo, EGYPT, p.p. 7, 22-26 Feb., (2003).

89- Alfred P. Wolf, and W. Barclay, Chemistry Department, BNL, Upton, NY (1973).

90- K. Debertin, and R.G. Helmer, "Gamma- and X-ray Spectrometry with Semiconductor Detectors". North Holland, Amsterdam, (1988).

91- G.F. Knoll, "Radiation Detection and Measurement", 2nd Edition. Wiley, New York, (1989).

92- G. Gilmore, and J.D. Hemingway, "Practical Gamma-Ray Spectrometry". Wiley, New York, (1995).

93- M. Sonck, J. Van Hoyweghen, and A. Hermanne, "Determination of the External Beam Energy of a Variable Energy Multiparticle Cyclotron", Appl. Radiat. Isot., Vol. 47, No. 4, pp. 445-449, (1996).

94- F. Tarkanyi, S. Takacs, K. Gul, A. Hermanne, M.G. Mustafa, M. Nortier, P. Oblozinsky, S.M. Qaim, B. Scholten, Yu.N. Shubin, Z. Youxiang. IAEA-TECDOC-1211, International Atomic Energy Agency, Vienna, Austria, p. 49, (2001). Available from http://iaeand.iaea.org.at/ medical, (2001).

95- F. Tarkanyi, F. Szelecsenyi, and P. Kopecky. "Cross Section Data for Proton, ^{3}He and α-Particle Induced Reactions on ^{nat}Ni, ^{nat}Cu, and ^{nat}Ti for Monitoring Beam Performance", Proc. Int. Conf. on Nuclear Data for Science and Technology (S. Qaim, Ed.), p. 529. Springer, Berlin, (1992).

96- P. Kopecky "Proton Beam Monitoring via the $^{nat}Cu(p.x)^{58}Co$, $^{63}Cu(p,2n)^{62}Zn$, and $^{65}Cu(p.n)^{65}Zn$ Reactions in Copper", Int. J. Appl. Radial. Isot. 36, 657, (1985).

97- F. Tarkanyi, F. Szelecsenyi, and P. Kopecky "Excitation Functions of Proton Induced Nuclear Reactions on ^{nat}Ni for Monitoring Beam Energy and Intensity", Int. J. Appl. Radial. Isot. 42, 513, (1991).

98- F. Tarkanyi, F. Szelecsenyi, and S. Takacs, Acta Radiol. Suppl. 376, 72, (1992).

99 K. Ishii, M. Valladon, C. S. Sastri, and J. L. Debrun, Nucl. Instr. Methods 153, 503, (1978).

100- Available from the International Atomic Energy Agency, nuclear data section, directory: physics\Ganaas Ver. 3.11, Vienna, Austria, (1995).

101- APTEC-MCA, verion 6.13, release 14, commercially available from Aptec Instruments Ltd., (1998).

102- I. A. Slavic, Nucl. Instr. and Meth., 134, 285, (1976).

103- W. Westmeier, Nucl. Instr. and Meth., 180, 205, (1981).

104- W. Westmeier, Nucl. Instr. and Meth., A 242, 437, (1986).

105- K. Debertin and U. Schotzig, Nuct. Instr. and Meth. 158, 471, (1979).

106- Tien-Ko Wang, I-Min Hou, and Chia-Lian Tseng, "Well-type HPGe Detector Absolute Efficiency Calibration and True-Coincidence Correction", Nucl. Instr. Meth. A 425, p.p. 504-515, (1999).

107- A. Wyttenbach, Coincidence Losses in Activation Analysis, J. of Radioanalytical Chemistry vol. 8, 335-343, (1971).

108- F. Cserpak, S. Sudar, J. Csikai, and S. M. Qaim, "Excitation Functions and Isomeric Cross Section Ratios of the $^{63}Cu(n,\alpha)^{60m,g}Co$, $^{65}Cu(n,\alpha)^{62m,g}Co$, and $^{60}Ni(n,p)^{60m,g}Co$ Processes from 6 to 15 MeV". Phys. Rev., C, Vol. 49, No. 3, (1994).
109- M. Gloris et al., "Proton Induced Production of Residual Radionuclides in Lead at Intermediate Energies". Nucl. Instr. and Meth., A 463, pp. 593-633, (2001).

110- E. Gilbert et al., "Cross Section for the Proton Induced Production of Krypton Isotopes from Rb, Sr, Y, and Zr for Enegries up to 1600 MeV". Nucl. Instr. and Meth. B, 52, pp. 293-319, (1998).
111- Photon Interaction Database, National Institute of Standards, and Technology, http://physics.nist.gov/phys, Reference Data, (2005).
112- Th. Schiekel, F. Subdrock, , U. Herpers, M. Gloris, H. J. Lange, I. Leya, R. Michel, B. Dittrich-Hannen, H. A. Synal, M. Suter, P. W. Kubik, M. Blann, D. Filges, Nucl. Instr. and Meth. B, 114, 557, (1996).
113- L. Moens et al., Nucl. Instr. and Meth. 187,451, (1981).
114- L. Moens and J. Hoste, Appl. Radiat. Isot. 34, 1085, (1983).
115- Tien-Ko Wang, Tzung-Hua Ying, Wei-Yang Mar, Chia-Lian Tseng, Chi-Hung Liao, and Mei-Ya Wang, "HPGe Detector True-Coincidence Correction for Extended Cylinder, and Marinelli-Beaker Sources". Nucl. Instr. and Meth. A 376,p.p. 192-202, (1996).
116- Sandor Sudar "TRUECOINC: A Program for Calculation of True Coincidence Correction for Gamma Rays", Institute of Experimental Physics, Kossuth University, Debrecen, Hungary, (2000).
117- R. Weinreich, H.J. Probst, S.M. Qaim, "Production of Chromium-48 for Applications in Life Sciences", Int. J. Appl. Radiat. Isot. 31, 223, (1980).
118- S.S. Malik, Nucl. Instr. Meth. 125, 45-52, (1975).
119- M.P. Fricke, W.M. Lopez, S.J. Friesenhahn, A.D.Carlson, D. Costello, "Measurements of Cross Sections for Radiative Capture of 1-keV to 1-MeV Neutrons by Mo, Rh, Gd, Ta, W, Re, Au, and U-238". Conference, 71KNOX, 1, 252, March, (1971); W.P. Poenitz, "Fast Neutron Capture Cross Section Measurement with the Argon National Laboratory Large Liquid Scintillator Tank". Symposium, ANL-83, 4, 239, April, (1982); A.R.DEL. Musgrove, B.J. Allen, J.W. Boldeman, R.L. Macklin, "Average Neutron Resonance Parameters and Radiative Capture Cross Sections for the Isotopes of Molybdenum". Nucl. Phys. A, 270, 108, Oct., (1976); J. Voignier, S. Joly, G. Grenier. "Capture Cross Sections and Gamma-Ray Spectra from the Interaction of 0.5 to 3 MeV Neutron with Nuclei in the Mass Range A = 63 to 209". Nucl. Sci. Eng., 93, 43, (1986).
120- W.P. Poenitz, "Fast Neutron Capture Cross Section Measurement with the Argon National Laboratory Large Liquid Scintillator Tank".

Symposium, ANL-83, 4, 239, April, (1982); R.C. Block, G.G. Slaughter, L.W. Weston, F.C. Vonderlage. "Neutron Radiative Capture Measurements Utilizing a Large Liquid Scintillator Detector at the Oak Ridge National Laboratory Fast Chopper". Conference, 61SACLAY, 203, July, (1961); J.H. Gibbons, R.L. Macklin, P.D. Miller, J.H. Neiler. "Average Radiative Capture Cross Section for 7 to 170 keV Neutrons Detector (STANK) Liquid Scintillator Tank". Phys. Rev., 122, 182, (1961); and Phys. Rev., 129, 2695, (1963); M.P. Fricke, W.M. Lopez, S.J. Friesenhahn, A.D. Carlson, D. Costello. "Measurements of Cross Sections for Radiative Capture of 1-keV to 1-MeV Neutrons by Mo, Rh, Gd, Ta, W, Re, Au, and U-238". Conference, 71KNOX, 1, 252, March, (1971).

121- Paul C. Brand, Materials Research Engineer, NIST Center for Neutron Research, Building 235, Gaithersburg, MD 20899, Interface between ICC and MCC, Revision - (DRAFT), (1999).

122- V.F. Weisskopf, and D.H. Ewing, "On the Yield of Nuclear Reaction with Heavy Elements". Phys. Rev. 57, 472, (1940).

123- M. Herman, A. Marcinkowski, and K. Stankiewicz, "Program Name: EMPIRE-MSC, Computer phys. Communicantion 33, pp. 373-398, (1984).

124- H. Feshbach, A.K. Kerman, and S. Koonin, "The Inelastic Scattering of Neutrons". Ann. Phys. 125, 429, (1980); R. Bonetti, M.B. Chadwick, P.E. Hodgson, B.V. Carlson, and M.S. Hussein, "The Feshbach-Kerman-Koonin Multistep Compound Reaction Theory". Phys. Rep. 202, 171, (1991); T. Tamura, and T. Udgawa, "Assessment of thc Feshback-Kerman-Koonin Appproximation in Multistep Direct Reaction Theories". Phys. Letts. B 78, 189, (1978).

125- E. Gadioli, and P.E. Hodgson, "Pre-equilibrium Nuclear Reactions", Oxford, Clarendon, (1992); V. Udgawa, and K.S. Low, "Assessment of the Feshbach-Kerman-Koonin Approximation in Multistep Direct Reaction Theories". Phys. Rev. C 28, 1033, (1983); K.K. Gudima, G.S. Mashnik, and V.D. Toneev, "Cascade Exciton Model on Nuclear Reactions". Nucl. Phys. A 401, 329, (1983).

126- M. Blann, "Hybrid Model for Pre-equilibrium Decay in Nuclear Reactions". Phys. Lett. 27, 337, (1971).

127- M. Blann, "Importance of the Nuclear Density Distribution on Pre-equilibrium Decay", Phys. Rev. Lett. 28, 757, (1972).

128- M. Blann, "Overlaid ALICE: a statistical model computer code including fission and pre-equilibrium models: FORTRAN, cross-sections". Report COO-3494-29. University of Rochester, NY, (1975).

129- M. Blann, Bisplinghoff, J., "Code ALICE/Livermore 82. Report UCID-19614. Lawrence Livermore National Laboratory, (1982).

130- M. Blann, Vonach, H.K., "Global Test of Modified Pre-compound Decay Models". Phys. Rev. C 28, 1475, (1983).

131- V.S. Ramamurthy, S.S. Kapoor, and S.K. Kataria, "Excitation Energy Dependence of Shell Effects on Nuclear Level Densities and Fission Fragment Anisotropies", Phys. Rev. Lett., 25, 386-390, (1970).

132- A.V. Ignatyuk, K.K. Istekov, and G.N. Smirenkin, Sov. J. Nucl. Phys., 29, 450, (1979).

133- C. Kalbach-Cline, "Residual two Body Matrix Elements for Pre-equilibrium Calculations", Nucl. Phys., A210, p.p. 590-604, (1973).

134- R.M. Lambrecht, T. Sekine, and H. Vera Ruiz, "Alice Prediction on Accelerator Production of Molybdenum-99", Appl. Radiat. Isot. 51, pp. 177-182, (1999).

135- M. Nandy, and P.K. Sarkar, "Calculation of Excitation Functions of Nuclides Produced in Proton Induced Reactions on ^{nat}Si", Appl. Radiat. Isot. 54, pp. 101-111, (2001).

136- A. Hohn, F.M. Nortier, B. Scholten, T.N. van der Walt, H.H. Coenen, and S.M. Qaim, "Excitation Functions of ^{125}Te(p,xn)-Reactions from their Respective Thresholds up to 100 MeV with Special Reference to the Production of ^{124}I", Appl. Radiat. Isot. 55 , pp. 149 -156, (2001).

137- M.C. Lagunas-Solar, Presented at The Consultants Meeting on Production Technologies for Molybdenum-99 and Technetium-99m, 10-12 April, Faure, South Africa, IAEA-TECDOC-1065, Vienna, Austria, (1999).

138- M.C. Lagunas-Solar, R.P. Haff, "Theoretical and Experimental Excitation Functions for Proton Induced Nuclear Reactions on Z=10 to Z=82 Target Nuclides". Radiochim. Acta 60, 55, (1993).

138- M.C. Lagunas-Solar, P.M. Kiefer, O.F. Carvacho, C.A. Lagunas, and Ya.Po. Cha, "Cyclotron Production of NCA ^{99m}Tc and ^{99}Mo, an Alternative Non-reactor Supply Source of Instant ^{99m}Tc and $^{99}Mo \rightarrow ^{99m}Tc$ Generators". Appl. Radiat. Isotopes 42, 643-657, (1991).

139- M.G. Mustafa, "Nuclear Modeling Applied to Radioisotope Production". Report UCRL-JC-126741. Lawrence Livermore National Laboratory, CA, p. 54, (1997).

140- P. Oblozinsky, "Development of Reference Charged Particle Cross Section Data-Base for Medical Radioisotope Production". Report INDC(NDS)-371. IAEA, Vienna, p. 45, (1997).

141- G.F. Steyn, S.J. Mills, F.M. Nortier, B.R.S. Simpson, and B.R. Meyer, Appl. Radiat. Isotopes 41, 315, (1990).

142- O. Bersillon, "SCAT2: Un programme de modele optique spherique". Report CEA-N-2227 NEANDC(FR), INDC(E)49, (1981); "The Computer Code SCAT2", NEA Data Bank 0829/03 version (1991).

143- J.J. Griffin, "Statistical Model of Intermediate Structure", Phys. Rev. Lett. 17, 478, (1966).

144- J. Raynal, "Notes on ECIS", CBA-N-2772, Commissariat a l'Energie Atomique (1994).

145- "Handbook for Calculations of Nuclear Reaction Data: Reference Input Parameter Library", IAEA-TECDOC-1034, 168 pages. See also www-nds.iaea.org/ripl/, IAEA, Vienna (1998).
146- W. Hauser, and H. Feshbach, "The Statistical Theory of Multistep Compound, and Direct Reactions". Phys. Rev. 87, 366, (1952).
147- H.M. Hofmann, J. Richert, J.W. Tepel, and H.A. Weidenmuller, "Direct Reactions and Hauser-Feshbach Theory", Ann. Phys. 90, 403, (1975).
148- H. Nishioka, J. J. M. Verbaarschot, H. A. Weidenmuller, and S. Yoshida, Ann. Phys., 172, 67, (1986).
149- V.V. Zerkin, "ZVView Graphics Software for Nuclear Data Analysis", Version 9.4, Available from: http://www-nds.iaea.org/ndspub/zvview/, (2001).
150- M.B. Chadwick, P.O. Young, D.C. George, and Y. Watanabe "Multiple Preequilibrium Emission in Feshbach-Kerman-Koonin Analyses", Phys. Rev. C 50, 996 (1994).
151- E. Betak, and P. Oblozinski, "Code PEGAS", Report INDC(SLK)-001, IAEA, Vienna, Austria, (1993).
152- M. Blann, "New Precompound Decay Model", Phys. Rev. C 54, 1341 (1996).
153- P.A. Axel, Phys. Rev. 126, 671, (1962).
154- A. Gilbert, and A.G.W. Cameron, Can. J. Phys. 43, 1446, (1965).
155- G. Audi, and A.H. Wapstra, Nucl. Phys. A 595, 409, (1995).
156- P. Moller, and J.R. Nix, Atom. Data Nucl. Data Tables 26, 165, (1981).
157- Mike Herman, Roparto Capote-Noy, Pavel Oblozinski, Andrej Trkov, and Victor Zerkin, "Recent Development of the Nuclear Reaction Code EMPIRE", J. Nucl. Sci. Tech., Suppl. 2, pp. 116-119, (Aug., 2002).
158- W.D. Myers, and W.J. Swiatecki, Ark. Fys., 36, 343, (1967).
159- M.M. Musthafa, B.P. Singh, M.G.V. Sankaracharyulu, H.D. Bhardwaj, and R. Prasad, "Semi-Classical and Quantum Mechanical Analysis of the Excitation Function for the ^{130}Te(p,n)^{130}I Reaction". Phys. Rev. C 52, (1995).
160- I.D. Fedorets. et al., Phys. Atom. Nucl. 56 (4), 436, (1993).
161- M.B. Chatterjee, Phys. Rev. C 2, 625, (1988).
162- J. Pal, et al., Z. Phys. A 356, 281, (1996).
163- T. Kibedi, et al., Phys. Rev. C 37, 6, 2391, (1988); G. Gulyas. et al., Nucl. Phys. A 506, 196, (1990).
164- P.E. Garrett. et al., Phys. Rev. Lett. 78 (24), 4545, (1997).
165- Mihaela Sin, "Empire-II Code System with RIPL Database as a Tool for Nuclear Spectroscopy", Nucl. Instr. Meth., A 527, p.p. 462-470, (2004).

166- D. Wilmore, P.E. Hodgson, Nucl. Phys. 55, 673, (1964); F.D. Becchetti, G.W. Greenless, Phys. 182, 1190, (1969); L. McFadden, G.R. Satchler, Phys. 84, 177, (1979).
167- V. McLane, C.L. Dunford, P.E. Rose, "Neutron Cross Sections", Vol.2, Academic Press, New York, (1988).

****

Printed by Printforce, United Kingdom